IMPLEMENTING OPEN SYSTEMS

The McGraw-Hill Series on Computer Communications

0–07–004674–3	Bates	Wireless Networked Communication
0–07–707794–6	Bearpark and Beevor	Computer Communications: A Business Perspective
0–07–709074–8	Bearpark	Protocols for Application Communication: A top-down guide to OSI, TCP/IP and SNA
0–07–005560–2	Black	TCP/IP and Related Protocols (2nd Edition)
0–07–015839–8	Davis and McGuffin	Wireless Local Area Networks
0–07–707883–7	Deniz	ISDN and its Application to LAN Interconnection
0–07–016751–6	Dhawan	FDDI
0–07–024037–X	Graham	The Networker's Practical Reference
0–07–042588–4	Minoli	Imaging in Corporate Environments
0–07–046461–8	Naugle	Network Protocal Handbook
0–07–707778–4	Perley	Migrating to Open Systems: Taming the Tiger
0–07–057442–1	Simonds	McGraw-Hill LAN Communications Handbook
0–07–060362–6	Spohn and McDysan	ATM: Theory and Applications

Related Topics

0-07-707881-0	Collin	The Essential LAN Sourcebook

IMPLEMENTING OPEN SYSTEMS

Daniel R. Perley

McGRAW-HILL BOOK COMPANY

London · New York · St Louis · San Francisco · Auckland
Bogota · Caracas · Lisbon · Madrid · Mexico · Milan
Montreal · New Delhi · Panama · Paris · San Juan · São Paulo
Singapore · Sydney · Tokyo · Toronto

Published by
McGraw-Hill Book Company Europe
Shoppenhangers Road, Maidenhead, Berkshire, SL6 2QL, England
Telephone 01628 23432
Fax 01628 35895

British Library Cataloguing in Publication Data

Perley, Daniel R.
Implementing Open Systems. – (Computer
Communications Series)
I. Title II. Series
004.680977311
ISBN 0-07-707948-5

Library of Congress Cataloging-in-Publication Data

Perley, Daniel R.,
Implementing Open Systems/Daniel R. Perley.
p. cm.
Includes index.
ISBN 0-07-707948-5
1. Computer architecture. 2. Open system interconnection.
I. Title.
QA76.9.A25P438
004.6–dc20 95-1106
CIP

1234 BL 9765

Typeset by Alden Multimedia Ltd
and printed and bound in Great Britain by Briddles Ltd, Guildford, Surrey.
Printed on permanent paper in compliance with ISO Standard 9706.

Contents

Trademarks

All computer vendor and product names used in this book, and the abbreviations thereof, are the registered trademarks of the respective vendors. For example, 'IBM' is the registered trademark of International Business Machines Corporation, 'DEC' is the registered trademark of Digital Equipment Corporation etc. Nothing in this book is to be taken as specific endorsement or criticism of any vendor or product.

Preface

Once you have begun the migration to open systems it is quite difficult to stop or to return whence you came, to proprietary systems. However, the journey is not always an easy one, and the taking of the initial few steps (although crucial) does not, in and of itself, guarantee arrival at the desired destination. Indeed, we are only just beginning to understand the nature and characteristics of that destination.

We do know that open systems offer tremendous new positive opportunities (which we can exploit) and pure benefits (which we will usually realize automatically). Open systems answer many questions that have been asked in past years, but they pose even more new ones. For example, If we can run the UNIX operating system on workstation, micro-based server, midrange system, midframe and mainframe platforms, it may be difficult to decide which platform on which to place a brand new application program. In the past, these things tended to be more obvious; if the application was too large for a PC it was placed on a minicomputer or a mainframe. In terms of how they interacted with the user, applications tended to be single-user, batch or multi-user (time-share). Today, we also have the client-server model and the demand for batch processing is falling rapidly in many organizations. So, application placement will become just as much an issue as the basic economic justification or the technical feasibility of a proposed project.

In my previous book, *Migrating to Open Systems: Taming the Tiger* some experience-based, and hopefully practical, advice was tendered as to how best to set about taming this wild beast called open systems. However, once you have managed to corral the tiger, the game is not over. Now you may wish to start *making the tiger dance*. While I resisted the temptation to add this phrase as a subtitle to this book, that is really the thrust of it. Now that you have successfully begun the process of in-migrating the first few open systems, you will soon have a community of them, and of open systems users. They will (quite rightly) expect IT management and staff to continue to demonstrate leadership when non-trivial issues such as application placement or system modelling arise. This book therefore addresses system modelling, application placement and a more formal version of cost–opportunity–benefit analysis. It seeks to tie the three of these together for reference purposes for readers and as a (foundation) functional specification for those who would automate all or part of the approach. The final chapter provides guidance on how best to utilize the planning methodology. Appendices provide background material supporting the methodology.

All of these should prove to be useful tools. However, they must be accompanied by a very important caveat. Of even the largest organizations, only a handful have had a significant amount of experience in making application placement and other such decisions in a *multi-tier open systems environment*. Therefore, most of the approach set out in this book is by definition more theoretical than it is extensively proven in the field. While it is based on my own practical experience – and that shared with me by many of my colleagues in this field over a number of years – it is not an authoritative text on the subject. It is, rather, a body of theory, some part(s) of which you may be able to select and put to good use. Naturally, it is my hope that the greater part of what follows will be of some assistance to you.

Daniel R. Perley

Acknowledgements

This book has proven to be a considerably greater challenge than my previous one, both because it tries to address a much wider set of issues, and because it is much more theoretical and conceptual. It has required an intense degree of concentration (and isolation) on my part and through all of this my family and friends have provided much (perhaps not always deserved) comfort, encouragement and support. This is both acknowledged and much appreciated. The patience of my business partners (through the lengthy and intense process of preparing this material) is also appreciated.

1 Taking stock of your position: a general introduction

Overview

This book is intended primarily for information technology (IT) practitioners who are now beginning, or have already begun, the process of migrating their organization's technology base to open systems. This exciting, yet often hazardous, undertaking offers more than its share of challenges and opportunities for professional (and personal) growth. My previous writings have concentrated largely on the generalization of lessons arising from my own experience, and that of a number of my colleagues, working in the initial stages of in-migrating new technologies. They provide a general approach, some major milestones and some (hopefully) helpful guidance along the way. This book is far more speculative, as it attempts to build a 'second storey' upon the foundation of open system architecture, which is itself the 'ground floor' of an IT program capable of acquiring and supporting both open system interconnection (OSI) networks as well as open computing systems (OCS). (For readers new to this field, these and other terms are defined in Chapter 2.)

The case for advanced concepts and planning tools

The arrival of the viable (versus the mere curiosity-value) form of open systems in the organization brings with it a number of new opportunities and challenges. Foremost among these is the need to be able to characterize the various technological elements in a manner which is both internally and externally consistent. This allows, among other things, dispassionate and logical comparison of the technological resources to be provided to two (competing) workgroups or the comparison of the current and planned technology base of a single workgroup. It also allows more definitive comparison either of the total degree of penetration of information technology within two organizations (such as prospective merger partners) or of the ratio of closed to open technology within each of them.

Another important function is the process of deciding where to actually place a new (or re-engineered/redeveloped) application program once it becomes available. In the past, these choices were quite simple, but such is not the case in a system architecture-driven OSI/OCS environment. Indeed, there are many situations in which dozens of factors must be considered in determining the appropriate platform on which to place a given application. With scalability of open systems platforms, some applications can run on any combination (or all) of notebook computer (NBC), personal computer (PC), workstation (WS), micro-based system (MBS), midrange system (MRS), midframe (MDF) or mainframe (MF).

Having linked open systems to business objectives (in their own minds and in the minds of senior management), some IT people may then believe that their economic (not to mention financial and accounting) work is done. However, such is not the case. A consistent economic model is required to enable case-by-case assessment and justification of exactly where and when to introduce new technology. To make the link between open systems and business objectives, and to delineate key positive opportunities and benefits, in the mind of your senior management is merely to convince them that the field is worth taking and that an army should now be committed to that task. To justify the development of each new application, and the acquisition each new item of equipment, is the actual 'trench warfare' of the process of migration to open systems.

Of course, you will never even reach the implementation and support stages if Department X deliberately entrenches, and thus stoutly resists your efforts to bring it into the age of the new technology. In many cases, the local manager will simply be ignorant of the true opportunities and benefits which open systems may present and thus (understandably) he or she may be somewhat resistant to change. This may be so even where senior management has trumpeted its commitment to the new technology from that high pinnacle called the boardroom. In other cases, locally entrenched IT professionals who report to the business unit's line manager (or else non-professional IT 'wannabes' who have appointed themselves as the local 'computer experts') may have galvanized the local line manager into resistance. Often these 'experts' have established a complicated local area network (LAN) with multiple servers which may even run different operating systems. In the worst case, your erstwhile proprietary vendor(s) may have decided that the best way to resist open systems is at the workgroup (trench) level by encouraging local line managers to retain what they now have or buy some new proprietary solution. Whatever the case, you need a good set of 'civil engineering' and 'war fighting' tools to allow you to take stock of the terrain and the opposition, build a logical plan of attack and execute it. The war analogy clearly has its limits, however, because this is not always a 'confrontational' situation. Nonetheless, in any situation where the new technology (and new way of doing things) is being introduced, it is necessary to be able to build a step-by-step chain of reasoning as to how to proceed. Just as important will be your ability to *explain* what you propose to do during the operation to the prospective patient!

There are, of course, some real dangers here which the reader must seriously consider before embarking upon consideration of the concepts put forward below. Few organizations have completed even the move to a full suite of OCS platforms and OSI networks. Fewer still have significant experience in managing large-scale re-engineering of *corporate* business processes co-incident with wholesale redevelopment of their core business applications. Most organizations have proceeded with open systems technology within, for example, one 'lead' theatre, be it the workplace tier, the workgroup tier, a specific department or region, or whatever. I have previously written that there is not yet enough experience anywhere in the world even to produce the authoritative experience-based work even on how to begin implementing open systems. That being the case, it can be readily understood that there is even less possibility of producing a rock-solid theoretical treatise on how one should proceed to conquer the entire field once a beachhead has been established. Therefore, by definition and by default, this book is at once more hypothetical and speculative than books which simply give an exposition of what open systems are all about or provide advice on how to start using them. Rather, the objective here is not to help you begin taming the tiger represented by open systems in the first place, but, rather, to provide some pointers on how you might make the tiger dance! Dancing with tigers may be even more exciting than taming them. Nor will the concepts in this book provide you with a Teflon suit. Therefore, you must judge (in your own organization's context) where these tools – or your own derivatives of them – may be of assistance in moving forward the cause of open systems, where they may simply muddy the waters and where they may actually be counter-productive.

Objectives of this book

To provide an analytical tool-set for open systems

While never a particularly quantitative sort of person by nature, my work has nonetheless sometimes forced me to take an analytic approach to open systems planning and implementation problems. Having had the opportunity – over the four-plus years since returning to the private sector from leadership of a large government open systems procurement program – to collect these experiences into some semblance of order, it has been possible to develop them into an interrelated and hopefully comprehensive tool-set for planning and implementing open systems, principally for the latter.

To allow you to recognize, quantify and exploit opportunities and benefits of open systems

My previous book contains an extensive menu of costs, opportunities and benefits which can be associated with open systems. A list of these is provided in Chapter 2 of this book. However, knowing that a mountain might contain (or even does contain) gold is not enough; one must have some idea of *how much* gold there is and of how to engineer its extraction, and must be able to calculate and compare the value of the former with the cost of the latter. Only then is it possible to know if there is the opportunity for a viable gold mine. Until you can make this 'business case', open systems implementation can never become a priority or a reality within your organization. If, however, you can (for example) show that there will be less frequent 're-inventing of the wheel' in local (branch or department) software development initiatives because application portability will indeed create a 'software commonwealth' (which nips useless parallel development in the bud) you can show a real saving. If this also involves more 'new' applications sharing the same platforms (versus each having their very own private platform), and hence fewer purchases of new platforms, you are well on your way to making a case for open systems. Being able to determine some sort of 'cost per pound' for original applications, for porting effort and for computing platforms is (unfortunately) essential to this process. This is unfortunate because computer technology is changing so quickly, and is so complex, that virtually any approach to this modelling and metricization can at best be criticized as arbitrary and at worst can be labelled quantitative quackery! Nevertheless, it is necessary to start somewhere.

To be intelligible and practical

Many people have provided positive feedback on the 'plain English translations' provided in my previous book. These have been continued here, but a caution is also warranted: senior management and other lay readers (those who are not IT professionals) must understand that this book addresses considerably more complex and esoteric issues and that my 'translation capability' is therefore at times strained almost to breaking point. Unlike, for example, German the English language does not suffer well the stringing together of many difficult concepts into single words (albeit very long ones in the German case).

What you will be able to do after reading this book

Determine when planning tools are required

Just as a soldier does not require a rifle (or even a bayonet) to open a foil-covered field biscuit, some of the analytic tasks facing open systems tiger-tamers will not require the tools provided here. This book will help you to decide when and where to use the various available tools.

Establish a planning framework

No matter how picky your accountants or how discerning your senior management, if the materials in this book cannot be put to good service to help render a bullet-proof plan (and justification) for migrating to open systems, then your organization has extremely unusual business and operational characteristics, its accounting/financial fraternity is getting kickbacks from a proprietary vendor or senior management has a death wish. Open systems are becoming essential to the survival of many organizations.

Model key open systems parameters and characteristics

The modelling approach put forward here is far from perfect and it may be superseded in the next few years by far more elegant, calibratable and effective models than my own. However, for the present, the proposed approach should allow you to take some measure of what you have in service now and to quantify some of the costs, opportunities and benefits in the context of your own situation.

Assess economic implications of alternative strategies

The beauty of any type of modelling or simulation is that you can use it to try out all sorts of 'what if?' questions and tactics without betting the IT shop (or even the whole organization's business) on an ill-fated experiment. We know, for example, from Communism, that when you use the whole country as an economic guinea pig the results can be disastrous. In the aviation world (and increasingly in surface transportation too), simulators are used to help design the operator–vehicle interface, train the operator, keep the operator proficient and retrospectively investigate operational incidents or accidents. Simulators of an open systems architecture can be similarly useful.

Determine optimum processing tier for an application

When a given application can run on anything from a glorified (intelligent) executive folio to a huge mainframe (since they all run UNIX) deciding which platform on which to actually place the application becomes a non-trivial question. It can certainly be argued that this process is both part art and part science; indeed, the method proposed in this book may be viewed by some as being too laborious or complex. However, my purpose will have been achieved if even part of it can be effectively used to help you gain a sense of how best to decide where to place new or redeveloped applications. Also, the IT model may assist in training your system architects and application developers to gain a good sense of how to let applications 'find their own level'; just like water. At the very least, this book identifies the range of issues you must *consider* when making application placements.

Assumptions made about the reader

Understands the basics of open systems

It is assumed that you would recognize an open system if you were to be presented with one. This ability may come from some combination of past experience, studying the (now rapidly expanding) literature and learning acquired from colleagues. A great number of books are now available on the subject of open systems. For those at the embarkation point, *Paradigm Shift: The New Promise of Information Technology* by Don Tapscott and Art Caston (McGraw-Hill, 1993) sets out how new technology and organizational reorientation and change are mutually reenforcing. In my own view, Pamela Gray's *Open Systems: A Business Strategy for the 1990s* (McGraw-Hill, 1991) and *Open Systems and IBM: Integration and Convergence* (McGraw-Hill, 1993) are both required reading for anyone seriously interested in open systems. John Quarterman's *UNIX, POSIX and Open Systems: The Open Standards Puzzle*

(Addison Wesley, 1993) is also a valuable contribution to the field. Lest the reader think the author to be biased in his own publisher's favour, it should be pointed out that there are a number of other books from other sources which cover similar ground, for example, Peter Judge's book *IT Standards Makers and their Standards* (Technology Appraisals, 1991).

My own previous book on this subject, called *Migrating to Open Systems: Taming the Tiger*, addresses not what open systems are, but rather how one might begin implementing them in the first instance. Chapter 2 of this book provides a summary of the core concepts presented in the previous one; however, it can cover these only at a high level. Therefore, while not a strict prerequisite, my previous book is (with an admitted bias on my part) recommended as a useful companion to this one.

Desires to move to open systems

It is also assumed that the reader, having digested some of the above-cited materials, is actually motivated to try to move his or her organization into the field of open systems. Some readers may be the senior management supporters of such an agent of change. The migration could already be in progress, although it is likely not complete. The tools provided here may assist in making this migration go more smoothly and/or more quickly than would otherwise be the case.

Recognizes that new planning tools are required

The reader must also recognize that the arrival of open systems changes most of the 'rules of the ball game' within the IT world. This is so within each organization, but naturally the local specifics will impact the results. Continuing to use one's hockey stick now that the game has changed to baseball is not very productive. No one ever hit a home run with a hockey stick!

Chapter structure

As stated above, Chapter 2 will provide the core or 'kernel' of the knowledge which is necessary to the successful implementation of open systems technology in most organizations. Chapter 3 proposes a System Functional Model (SFM) which models how all installed IT infrastructures, both open and closed proprietary ones, behave operationally. Chapter 4 sets out a comprehensive Economic Model of Information

Technology (EMIT) which combines the underlying economic issues and the SFM to create the means of characterizing the 'costs of doing business' in various IT regimes. Chapter 5 proposes an Application Placement Methodology (APM) which, being aptly named and to no one's surprise, is the part of our model which purports to be of some assistance in deciding where to place applications. Chapter 6 draws the previous three chapters together into a coherent whole called the Planning and Implementation Model of Open Computing Systems (PIMOCS). In conclusion, Chapter 7 attempts to equip the reader with realistic expectations of a planning methodology and may help to modify the model to meet his or her own organization's requirements.

2 Open systems: origins and a good foundation

Introduction

This chapter presents some of the key concepts related to the process of migration to open systems. It is based on my previous book, *Migrating to Open Systems: Taming the Tiger* and is intended for readers unfamiliar with that material. It does not provide a full treatment of the subject, but rather a high-level overview. Readers unfamiliar with even the basics of open systems may wish to consult Pamela Gray's book, *Open Systems: A Business Strategy for the 1990s* or another similar introductory text.

The challenge: taming the wild beast

Despite recent progress, there remain many organizations wherein the only open systems are those being read about by the IT staff (or end-users) in magazines. This chapter concentrates on a few of the more vexing elements of actually migrating to open systems from an IT manager's perspective. These include obtaining senior management support, establishing a coherent technical approach, assisting the IT staff to adjust to the change and arranging measures to keep up with the technology. We therefore consider the early, and latter, stages of the migration process while not addressing actual trials, pilots and productionization.

Anyone who ever thought that a move to open systems (within a large and established organization) is easy probably believes that an ocean liner can turn on a dime. Putting the requisite measures in place to help a large private firm, institution or government department move to open systems can be a considerable challenge. Planning for, and acquiring, open systems is much more complex than for proprietary systems such as VAX/VMS or even the widely used DOS/Intel standard. This not only leads to higher front-end costs for most elements, except the hardware and software themselves, but also requires the IT organization first to reorient and retrain its own staff and end-users. It must then decide whether to make or to buy far more of the services previously provided by the proprietary vendors.

Closed systems and open systems

Over time, all of the computer vendors developed their own processors and communications technologies, with their related low-level languages and protocols, which in turn gave rise to varying operating systems and high-level languages. These differences, plus additional real or perceived value added by each of the vendors, tended to make commitment to one vendor's products (or even commitment to one product line or architecture within a given vendor's overall offering) a very long-term proposition. As more and more investment was made in training, operating experience, system software modifications and development of applications software, and in more terminals, printers and other peripherals, it became very difficult for a user organization even to consider changing products or vendors.

☛ Computer makers each developed different hardware and operating systems. Therefore, their languages (and of course the programs written with such languages and the data they handled) could only be used on their own machines. The customer becomes 'locked in' to one vendor; this is great for the vendor, but can be costly and troublesome for the customer.

Major user organizations came to believe that it would be desirable to remove the barriers that stopped them from mixing and matching components from different vendors. The only logical way to do this was to find a way to force the components of the computer system to adhere to standards. Standards are normally defined in terms of any or all of form, fit and function. For example, in North America, VHS and Beta have become the standard videotape formats, and in Britain everyone drives on the left side of the road.

An open system, as defined by the IEEE POSIX committee in the Guide to the POSIX Open Systems Environment is:

> A system that implements sufficient open specifications for interfaces, services and supporting formats to enable properly engineered applications software:
>
> - to be ported across a wide range of systems (with minimal changes)
> - to inter-operate with other applications on local and remote systems, and
> - to interact with users in a style which facilitates user portability

This is a very useful definition. However, I do not believe that it goes far enough. In reality, the term 'open systems' refers to two very distinct sets of standards, products, features and characteristics, as defined below:

1. Open systems interconnection (OSI) is a seven-layer hierarchy of communications standards, conventions and protocols which permit interconnection, and in some cases inter-operation, of similar or dissimilar computer systems. These standards are set and managed by international standards bodies and are the basis for international, industry-wide efforts to standardize the ways in which automated systems communicate with each other. Note that the TCP/IP (actually the whole Internet Protocol Suite (IPS)) also provides similar benefits and is more mature than pure OSI.

☛ OSI relates to standardizing how computers communicate with each other. It is a means of making similar or dissimilar computers use a common approach to becoming aware of each other, having a dialogue, exchanging information or working together. This common approach is defined in terms of a hierarchy of mandatory and optional ways of packaging and sending signals between the computers. Referring to the five-layer stack above, it lets computers with totally different layers communicate with each other.

2. Open computing systems (OCS) are computer systems which, through the use of common or compatible components such as system software (operating system), relational database management systems (RDBMS), programming languages and other components, permit vertical, horizontal and forward portability of user environments, applications, data and packages. Open systems exhibit a high degree of application, data and user environment *portability*, *scalability* and *interoperability* with each other.

☛ OCS relates largely (but not entirely) to standardizing the most basic programs that computers have, the programs that link the computer itself to the higher level programs that actually do the work for the user. These basic programs are called operating systems, often also known as system software. By running the same operating system on two very different computers, it becomes possible to transfer (port) a high level program from one type of computer to another. Having the same (a standard) operating system on two different computers helps make things more open, although sometimes it is not enough. Using the same computer programming language also helps, as does using other software which runs on both computers. With OCS you can carry a user's program to a larger or smaller computer, as well as to a replacement computer from a different vendor, much more easily. This is called three-axis portability: vertical, horizontal and forward.

The hyper-complexity of inducing computers to communicate with each other *within* companies and countries, much less among them, caused the evolution of a new set of communications standards and approaches, broadly labelled the open systems interconnection (OSI) approach. The International Standards Organization (ISO), based in Geneva, Switzerland, which coordinates the activities of many national standards bodies working in every field from screw threads to road construction, began to focus on the desperate need for standardization in the computer field. Standards would allow diverse computers not only to be connected together but to actually communicate and work together (to 'inter-operate'), would promote the distribution of applications (to bring them close to the users by having the same program run on small computers that formerly ran only on large ones) and would make computers and communications networks easier to manage. In this sense, OSI is a distinct concept, different from the operating system or other commonality; although it promotes inter-operability, it does not allow application programs to be ported from one computer to another.

A seven layer model of computer communications, over a network, was created as follows:

- The PHYSICAL layer handles the encoding of data into signals able to be sent over mechanical, electrical, radio channel and/or fibre transmission media
- The DATA LINK layer synchronizes transmission and detects/corrects errors
- The NETWORK layer establishes, maintains and ends communications between two nodes (and their respective computers) by establishing a logical path
- The TRANSPORT layer gives the user or application the 'handles' needed to control the functions of the previous three layers
- The SESSION layer oversees the dialogue between the computing systems such that each computer sends information when the other is expecting it
- The PRESENTATION layer converts data for presentation to the recipient computer in a form it will understand
- The APPLICATION layer provides specific services which complement efforts by the applications or end-users to transfer files or to access each other's databases

☛ This is the hierarchy of approaches to packaging (nesting) and sending signals between computers which was referred to in the last English translation section. It is becoming very widely accepted in Europe, in North America and elsewhere.

The above is also called the 'OSI Reference Model'. Taken together, these seven layers provide a framework or model for 'slotting in' individual communications facilities and services provided by a computer, a communications node (which itself may be a form of computer) or by a private or public network. The model is vendor-independent, letting even two proprietary and otherwise incompatible computers at least communicate with each other. It can do even more when these computers are themselves both open systems, as defined above. OSI has had the problem of slow and tortuous implementation, and a networking standard called TCP/IP has been widely used in the interim, particularly in North America. Many people believe that TCP/IP might even supplant OSI in the short to medium term, simply because it is so widely used, and because many communications components located higher up the OSI stack can run over a complex of lower layers made up of TCP/IP protocols.

The term 'UNIX' is used in this book as a generic description of any open operating system (one which runs on multiple vendors AND multiple architectures) and includes AT&T UNIX V.3, Berkeley BSD, XENIX, SUN-OS, UNIX International UNIX V.4, Open Software Foundation OSF Version 1 and a number of others. Almost all UNIX and UNIX-like operating systems trace their heritage back to work performed at AT&T's Bell Laboratories. Very recently, the X/Open Company Ltd. has become the holder of the UNIX trademark, which will eventually apply only to those operating systems which X/Open has itself branded. The branding process forces vendors to meet rigorous technical and interface requirements and to resolve inconsistencies which may arise between/among their products at their own cost and expense, rather than dumping the problem on the customer.

☛ Operating systems called 'UNIX' (and those called generally similar names, usually ending in 'IX' and 'UX') permit programs to be carried more easily from one computer to another.

The term 'POSIX' defines a family of standards developed by a group of committees under IEEE auspices which specify operating system and other *functional and user interfaces* but which do not specify, for example, the operating system itself. POSIX is important to portability but does not, on its own, guarantee portability.

☛ Do not be fooled – POSIX is *not* an operating system! POSIX defines that way in which users (and certain elements of the computer itself) will *interact with* the operating system, but says less about the operating system itself. Proprietary operating systems can become POSIX-compliant. That does not make them into open operating systems; in some cases it merely gives them a 'cloak of openness'.

A notebook computer (NBC) is a DOS/Intel, MAC OS, OS/2 or UNIX-based small and highly portable processing system, usually including an integral CPU, folding screen and compact keyboard and often incorporating a trackball or stowable mouse pointing device. The NBC will include the same types of ports for peripherals and communications as a PC, described below.

☛ There is a wide range of notebook and similar computers now available.

A personal computer (PC) is a DOS/Intel standard-based processing system running the MS-DOS operating system, usually including a central processing unit (CPU) cabinet, a keyboard and a cathode ray tube (CRT) screen. In certain cases, OS/2 or even Windows NT may be offered as an alternative operating system running on the Intel CPU platform.

Often a printer and mouse accompany the system. MS-DOS is technically a proprietary operating system (although DR-DOS has effectively replicated it) since it is controlled largely by one vendor and only runs on one computer architecture (Intel's 8086, 80286, 80386, 80486 and successor processors). However, the DOS/Intel combination is so now widely used that it behaves something like an open operating system. OS/2, in contrast, is an almost entirely proprietary operating system. There is avid price competition among suppliers of hardware and software. The PC is by far the most successful component of the microelectronics revolution; it has gone a long way towards democratizing the technology, placing it in the hands of millions of users.

☛ Most desktop computers are of the PC genus.

The term 'workstation' (WS) is usually used to describe a more powerful personal computing system. Many workstations run UNIX, but others run the DEC VMS or other operating systems. At any given time, the tendency has been to use the term WS to describe a desktop system more powerful than the then-current PC. This has become very confusing since, for example, today's 80386, 80486 and Pentium-80586 based PCs are more powerful than the power-user 'workstations' of just a few years ago.

☛ PC and WS equipment is for a single user and normally sits on or near the desktop, although the same processing engines used to drive these systems are found in everything from factory robots to the cars they build.

A micro-based server (MBS) is a system based on PC technology which is able to control a local area network (LAN) and provide access to data, communications services and certain programs to end-users with PC or WS equipment. Usually, the MBS and the PC or WS function as a team. The MBS, which has more storage capability, runs part of a program and the desktop device runs the other part. The part on the MBS usually finds the required data (and sometimes a copy of the program logic needed to manipulate that data) and 'serves' it the PC or WS. The latter uses its own processing power to perform the required functions on the data.

☛ An MBS functioning as a server can be thought of as an over-grown PC that acts as a sort of a 'mother hen' to a number of other PCs, passing them what they need to do their respective jobs and then later taking back the results for safe-keeping. While an MBS does this job very well, when the data set grows too large, the logic becomes too complicated or the number of PCs in the brood becomes too many, problems will result. Then it is necessary to choose between having more than one MBS or moving to a larger system, as defined below.

A midrange system (MRS) is a multi-user computer which is installed in an office environment to support a workgroup with anything from five or ten to approximately 70 members. It can facilitate access to a mainframe, can run office automation (OA) software and a relational database management system (RDBMS), can run applications written in third-generation languages (such as C or COBOL), supports PC internetworking and can function as a LAN server.

☛ The MRS is the current counterpart to the traditional 'minicomputer', but there are some important differences. Unlike the minicomputer, most MRS equipment does *not* require a special glass house or even a dedicated room in which to live. It can use the standard office (or even home) 110 V or 240 V A.C. power and its sound/heat output are compatible with placing it in someone's office or cubicle. Another important difference is that today's MRS equipment can be taken care of (administered) by a para-professional trained and certified for this purpose. It does not need a computer expert to look after it on a day-to-day basis.

The terms '*server*', '*LAN server*' or '*file server*' apply to any device (be it a dedicated standalone MBS or an MRS) which is functioning as the controller and file server for a LAN. (Note that some LANs may have more than one file server, but will have only one controller.) This server receives requests from intelligent (PC or workstation) clients and responds to them, usually transferring a copy of the requested file to the client and receiving an updated copy back from the client later. The relationship between 'LAN server' and 'MRS' is not reciprocal: while virtually any MRS can be equipped to function also as a LAN server, not all LAN servers can be equipped to function as an MRS. Most MBS class systems can only time-share five or at most ten users, providing such users are not overly demanding in terms of processor capacity. Many MBS-based LAN servers have insufficient CPU power, storage or I/O capacity to serve as a true multi-user MRS. While a LAN server is primarily a file server it may also provide client access to communications services (including gateways to networks and mainframes) and single-user RDBMS or other single-user applications. Applying the term 'RDBMS server' or 'application server' to an MRS is very often incorrect, since an MRS normally runs a multi-user versions of the RDBMS, RDBMS-dependent applications and other applications; in this case the RDBMS or application is not 'served' to an intelligent client but the user merely accesses it as a terminal or emulated terminal on a PC. Likewise, calling an MRS a 'communications server' is incorrect, because the server model implies intelligence at the client (user) location and single-user server-based communications applications, whereas an MRS can provide proprietary terminal and controller emulation (for mainframe access) to one or more real (or emulated) non-intelligent ASCII terminal-equipped users.

A client-server service model is one in which a server, as defined above, receives requests from intelligent clients and responds to them, usually transferring a copy of the requested file to the client and receiving an updated copy back from the client later.

☛ Here, each of the MBS or MRS (or even a large mainframe system) and the PC do part of the work. The work is divided according to its functional content.

A multi-user service model is one in which an MRS, functioning as a true multi-user system, permits two or more users with actual or emulated terminals to make simultaneous use of the same application or package. Locking at the file and/or record level prevents two or more users from simultaneously attempting to modify the same data. The operating system manages all user interactions and processes.

☛ Here, the MRS or larger system does all or virtually all of the work; the PC or WS may package and present data for the user's inspection, however.

Client-server and timesharing service models are not mutually exclusive: an MRS can be equipped to facilitate either or both of them. The only absolute restrictions are that client-server architecture requires the users (clients) to have intelligent workstations (or at least quasi-intelligent terminals) and client-server is also not truly multi-user (because the server deals with each client in isolation from all other clients, as if they did not exist).

An X-terminal (XT) is a quasi-intelligent reincarnation of the old non-intelligent terminal. It can do as good a job as a PC by 'fooling' an MBS or MRS into believing that it is a fully intelligent 'client', able to do cooperative two-level processing, as defined above.

Open systems and business objectives

As was stated in 1990 by the User Alliance for Open Systems, it is necessary to link open systems to business objectives to determine the leverage available from implementation. The portability, scalability and interoperability (PSI) 'benefits' of open systems are those most often cited. However, a wider view is necessary to present the pros and cons of open systems honestly to senior management, in the context of the specific organization's situation. Also, traditional financial analysis and cost–benefit analysis is not altogether useful for determining when and how to move to open systems, since most of the costs are up-front or at least near-term, and many significant benefits are longer term. Also, many of the costs are infrastructural or indirect.

More importantly, open systems offer many things to an organization which can neither be characterized strictly as benefits nor strictly as costs. These are 'opportunities' in that they offer the chance to obtain something good from open systems; however, many of them – if neglected – will come around and bite us in the posterior as surely as an inattentively managed tiger. Security is an excellent example; for a number of years open systems were widely maligned for being 'less secure' than their proprietary counterparts, particularly in the world of mid-range systems (MRS). While a proprietary MRS may have better 'out of the box' security, open systems contain all of the basic building blocks permitting a system to be made as secure as necessary, no matter what the identified threats. The difference is that the purchaser cannot take security for granted; it is necessary to actively *make* the system more secure. Failure to do so will likely leave it less secure than a 'stock' proprietary system.

In studying the business case for an open system, far more than for a proprietary system, one must make full recourse to the dynamics of, and leverage available from, the organization's own business environment and business mission in seeking to establish short-, medium- and long-term costs, opportunities and benefits of open systems. Failure to exploit the available 'opportunities' will at best cause your estimate to understate the positive aspects of such migration and will at worst create the opportunity for operational disaster. Therefore, the implications of adopting open systems for a given organization must be seen in terms of:

- COSTS, which are purely financial costs or disadvantages, not displaying any positive aspects in and of themselves
- OPPORTUNITIES, which are situations or circumstances created by the adoption of open systems with potential outcomes ranging from positive to negative
- BENEFITS, which are purely operational benefits or financial savings, not displaying any negative aspects in and of themselves

Thus, costs are purely negative or disruptive, while benefits are purely positive or facilitative. Opportunities are those things which are not purely the one or the other, but which must be 'netted out' within themselves before being classified. In some cases, the positive and negative aspects of an opportunity can be reduced to internal costs and benefits, within the sole context of that opportunity, and netted out in financial terms. In others, such as security, a purely monetary orientation is not useful.

There is a tactical and a strategic realm for each of the costs, opportunities and benefits of open systems; the divisions presented in the accompanying lists are somewhat arbitrary but are based on real-world experience (Fig. 2.1). Table 2.1 can be used to determine which tactical and strategic costs of open systems would be most relevant to your organization. Most organizations of greater than a few hundred persons

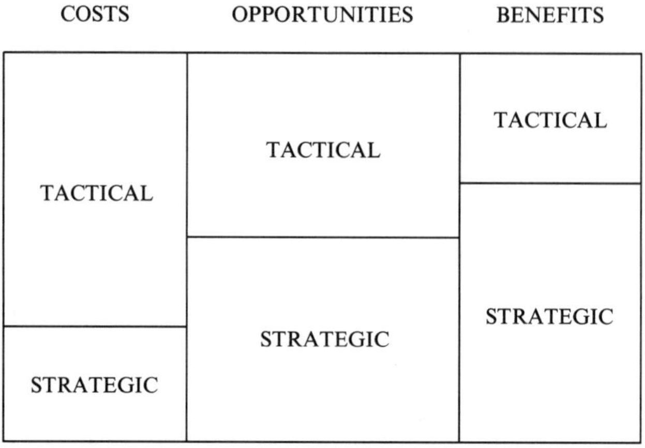

COSTS OPPORTUNITIES BENEFITS

TACTICAL TACTICAL TACTICAL

STRATEGIC STRATEGIC STRATEGIC

Figure 2.1 Open systems – costs, opportunities and benefits.

will incur almost all of them. Table 2.2 provides a virtual 'gold mine' of opportunities; you will only gain from an opportunity which is both relevant to your organization *and which you actively exploit*. Table 2.3 lists many of the (almost) automatic benefits of open systems; here again, though, not all of the benefits may be relevant; eliminate those that will not assist you! Therefore, your organization-specific 'C-O-B profile' consists of all the listed costs (less any you are *certain* you can avoid), those opportunities that are both relevant and which you actually intend to seek, plus all benefits except those which obviously do not apply to you. Thus, prepared, you have a much better chance of doing two things:

- linking open systems to business objectives in a manner which prevents would-be detractors from un-linking them
- developing an airtight business case for open systems which actually rank-orders the positively exploitable opportunities and pure benefits according to the prioritization of your organization's own business objectives.

(These accomplishments cannot fail to impress your top management.)

Developing a strategic plan for open systems

The plan should first address your kind of business, your business mission and the specific business objectives which your adoption of open

Table 2.1 Pure costs of open systems

Tactical costs	Strategic costs
Hardware	Architecture planning and policy/standards development
System software	Planning tools
Package software	Pilot projects and trials
Application software	Acquisition programs
System integration/commissioning	Operational costs
Orientation and training	Technology tracking: coping with the information boondoggle
Software/hardware support	Loss of one-stop-shopping
System administration	Conversion of applications and data
OSI communications overhead	UNIX Technology lag
Foregone features	Protracted proprietary vendor misbehaviour
Early retirement of existing proprietary system	Early bird standard selection syndrome
Incremental liability insurance (untried system)	Software vendor lock-in potential
OSI product overpricing	Requirement to restructure data to relational model
Increased failure risk	Premature discard of interim solution
Lower functionality at the outset	
Longer implementation and learning period	
Re-education/reorientation of your vendor	
Immature UNIX mainframe cost and disruption	
User frustration	

Table 2.2 Opportunities offered by open systems

Tactical opportunities	Strategic opportunities
Improve individual and workgroup efficiency and effectiveness	Three-axis portability (horizontal, vertical, forward)
Freed-up time available for creativity, QA or additional production	Scalability
Potential for inter-tier processing trade-offs	Interoperability
Work, work item (WI) and workspace sharing	Unified configuration management
Purchase decisions made on price/performance basis	Remote system administration
DOS interworking potential	Easy upgrade for symmetric multiprocessing
Lower system integration/support costs (after learning curve)	Hone sourcing strategy to architectural component
Progressivity UNIX licensing costs	Support the home workplace
Potential for enhanced security	Develop applications once for all platforms
Qualitative and ergonomic/health impacts	Network expandability
Maintain or enhance competitiveness	Increased potential for software re-engineering
Maximize capability to exploit expansion opportunities	Unlock corporate information base to all users
Maximize flexibility and resilience in conditions of change	Match technology supplied to user needs
Reduce response time	Wide access to open systems skills
	Use same packages at all processing tiers
	Application placement flexibility
	Adopt system architecture driven approach
	Framework for future expansion

Access more applications (more than all but DOS/Intel)	Allow organization to be business-driven (vs. technology-driven)
Provide application developer best possible environment	Improve organizational mergeability
Balance developer creativity and discipline	Improve downsizability
Utilize OSI to unite existing proprietary environments	Cut IT response time (to user requests)
Faster introduction of new information technology products	Adopt common GUI
Faster introduction of new technology related to organization's business area	Align OSI, UNIX and application development strategy
Provide system facilities not otherwise available	Create customized interoperability environment
Vendor-independent transaction processing	Achieve consistency in standard/profile selection
Integrate GUIs	Treat information as an asset
Better information exchange with customers	IT concentrates on user problems
	Provide UNIX programming benefits
	Evolve to knowledge-based organization
	Change organizational culture
	Better alignment of business and IT objectives
	Consolidate networks
	Optimize pace of technology adoption

Table 2.3 Pure benefits of open systems

Tactical benefits	Strategic benefits
Wider choice of business applications	Lower network investment
Integratable office (workplace) automation	UNIX has inherent network orientation
PC support by local system administrators	Vendor independence = corporate life insurance
Increased access to mainframe	Multi-vendor networking
Future disruption avoidance	Separate communications service and facility purchasing
Improved price/performance (net installed benefit)	Avoid maintaining geriatric applications
Lower LAN administration costs (servers can grow)	Improve leverage due to knowledge-based organization
UNIX/OSI/Ethernet synergies	Improve senior management visibility/control of business processes
Better external communications access	Improve IT staff efficiency and effectiveness
Faster intra-organizational communications	Achieve IT predictability for end-users
Low-cost data transfer	Increase flow of new product releases available
Avoid manual re-input of data	Acquire products in competitive market
Improved communications facility utilization	Increase product/version installed life
Shared workspace with heterogeneous systems	Distributed development of application components
Improve staff transferability	Provide any amount of processing power, anywhere
Reduce software support/upgrade costs	Reduce number of networks and operating systems
Lower external support costs	
Better use of heterogeneous communications	
Support autonomous workstation users	
Multiple applications share data	
Collaborative application development (among organizations)	

systems is intended to help achieve. It should discuss the types of information managed in your organization and highlight those that will soon (and later on) be impacted by the new systems. Then it should discuss the current (and anticipated future) business applications which will manage such information, followed by the packages (RDBMS or whatever else) upon which those applications will depend. Then, and only then, should the plan discuss the types of computer technology which it is anticipated will be needed to support the rest of this hierarchy. It should do so only in general terms. At this stage it is quite acceptable to say that we want to move most logic processing to the workgroup level. It might be premature to state that we will use only MBS or only MRS equipment, or both. It would certainly be premature to state here that we will acquire computers of Brand A and Brand B only.

Putting plan into practice

System architecture

Without a system architecture, you will not even know if a given new product you see at a trade show could even fit into your plans, much less where to put it. From the strategic plan, and from your body of information on user requirements (perhaps after appropriate studies in this area), you should have a system architecture produced which addresses at least the physical and either the logical or functional views.

☛ You have to have a map and plan which tells you where each piece of technology fits in and how it will work with those adjacent to it in the structure.

Central technology management infrastructure

This is the time to establish a skeleton structure, perhaps even the actual organization to manage the inflow and support of the new technology. There are two schools of thought on this: one is to develop the capability totally within your existing structure, so it is integral from the outset, while the other is to establish a discrete unit. Each strategy has its advantages.

Information management (IM) planning

If your organization has not already implemented (or purchased from outside consultants) an IM planning capability, you will want to delay no longer in this regard. A good number of the opportunities and benefits of open systems relate to better management of information as a corporate resource. If this concept has not yet been introduced into your corporate culture, you may wish to proceed immediately to introduce it. On the other hand, you may already have a comprehensive IM plan in place. In any event, you will want to ensure that someone is responsible for regular updating of the plan.

☛ You need someone to track what information your organization acquires, produces, uses and distributes. When you know the creation and flow of information, it becomes easier to conceptualize business applications which manipulate such information and to visualize how they will work together.

Current and planned application portfolio

Once you have a basic understanding of what information is being managed by whom, where and in what ways, you can begin to address the application architecture component of system architecture in more detail. What portion of the information will be managed by old applications in new clothes? How many can be migrated from legacy systems to open systems? How portable are these old applications likely to be? How much re-engineering will be involved? For new applications, what will be the split among in-house development, contracted development, third-party vendor and recycling as sources for applications for various purposes? Will you adopt standards and form-fit-function specifications for OA and RDBMS products, or will you actually certify one (or more) products for each of these roles? In given circumstances, how will you decide where to source an end-user business application?

☛ Knowing your current and future inventory of the computer programs that actually do the work provides a good idea of how much technology you will require. In subsequent chapters, we will consider how to decide how much computing power to use to support a given application and (the flip side issue) of what scale of platform on which to place an application.

Understanding and tracking the technology

Whether or not you establish a discrete organization to lead the move to open systems, you will need people to track various technologies of interest on an ongoing basis. The 'watch and brief' people are a good start. Some of them may be candidates for elevation to more formal technology monitoring responsibilities, while others may lose track of the ball, either accidentally or on purpose. Still others may decide that your current dominant vendor is the only source of information on open systems worth bothering with. If you are going to bring in new people, this is one of the best areas to do it, since it is just not possible to have too much expertise applied to this area. The more knowledgeable you are, the better will be the questions you can ask any presenting vendor who inundates you with claims of their long-standing tradition of openness.

☛ You will need people to track the technology so that you can anticipate and deal with changes in what is available to you in the market *before* they occur, not after.

System integration

You cannot move to open systems without having the capability to at least oversee and manage the integration of diverse products from various vendors. It may or may not be necessary to have the in-house capability to perform this integration, but you *must* be able to tell when the process is finished and the system is integrated. Over the past few years a number of organizations in North America and elsewhere have become experts in providing system integration services to their customers. This has usually been very helpful to the IT organization, because it permits one-stop shopping without necessarily committing to one-vendor shopping. Indeed, you can have an external system integrator actually integrate any combination of products, sourced in any number of ways by the integrator's organization or your own. You have a single point of accountability (the integrator is responsible to you for a coherent working system), but you need to be able to settle on a contractual (and if necessary legally enforceable) description of just what that is.

Many IT organizations have developed very considerable system integration capabilities in-house. Where such capabilities exist, you will almost certainly want to make good use of them during the tentative, early stages of your move to open systems. This should reduce by one the number of things that can go wrong. Later, if you intend to acquire a large number of systems (particularly if they are to be MBS or larger) in a relatively short period of time, you may need to supplement your own people with external assistance.

☛ Someone has to put the pieces together in a coherent fashion so that they work together to run the application smoothly, accomplish the business objective of the application and satisfy the users.

System administration and management

Open systems permit you to democratize much more of the computing power in your organization, to take it out much closer to the real users.

Unless you intend to stick to a purely mainframe and PC architecture as you move to open systems, MBS and/or MRS equipment *will* be introduced. When workgroup equipment is introduced you must think in terms of supporting workgroups. Actually, you have several choices.

1. Place an IT professional at each workgroup site, or at least each facility where there are a few workgroups using such equipment. This person can be a relatively junior support specialist and can certainly be trained to handle all of the required MRS, MBS and LAN administration as well as that required for the OA and RDBMS products.
2. You can buy what are sometimes colloquially referred to as 'care and feeding' services from a firm specializing in a mix of remote and on-site system administration. These well-trained tiger feeders will keep your 'beast' well cared for and under reasonably tight control. They can also support PC users where desired. Of course, they may not be as familiar with all of your applications as you would like, and may send a different person on every second visit. They are expensive, although you only pay for what you actually use.
3. Today, the technology exists to permit you to conduct virtually all system management functions (configuration management, capacity planning etc.) and a significant portion of system administration functions remotely, from a regional data centre or even from your central IT facility or corporate data centre. This approach usually requires not only that you have an implemented OSI (or at least TCP/IP capable) network in place, not a vendor proprietary network, and also that you have a very proactive approach to network management. Also, you will still require someone on the ground at the site to assist with any matters you simply cannot control remotely.
4. You can train (and certify) para-professional administrators of MBS/LAN and MRS installations. The term 'para-professional' is used here only as it relates to the IT field. A LAN administrator (LANA) or MRS administrator (MRSA) may well be a professional in another field, such as an engineer or a chartered accountant. In any event, this person will be trained and certified to do several things:

   - administer the workgroup system
   - provide basic support to users for PC and MRS operating and application software
   - know when they are 'out of their depth' and also who to call for help (IT, vendors etc.) in what circumstances

☛ There are several ways of providing technical and user support to the workgroups who use open systems. Which one you choose is less important than the fact that you *must* choose one. If you do not, you are abdicating your responsibility as leader of the transition to the new technology. Also, you will be steering your ship directly into a field of icebergs.

Networks and network management

You may already operate one network or a number of networks; for the moment any existing network(s) will be assumed to be proprietary. Most proprietary networks have at least some capability to allow you to interface other vendors' proprietary and open systems with them, however the level of interface is typically quite low and the 'outsider' equipment (unless it is of the 'plug-compatible' variety) is almost always treated as a second- or third-class citizen. In many cases, it will be preferable to use MRS or MBS equipment to emulate the unique proprietary terminals, terminal controllers and other vendor-specific gear which is now interposed between the end-user and your proprietary mainframe.

☛ You can use a UNIX workgroup system to fool your old mainframe into thinking it is still talking to its own terminals when it is really talking to users with many different kinds of terminals and PCs.

Optimum sourcing strategies

Sourcing strategies will vary almost as much as do organizations, but here the key issue is the *number* of vendor sources to be selected and ratified for a particular component of the system architecture. The three basic choices are set out below.

☛ You have to figure out how many different vendors' products you would like to have for each slot in your system architecture. Then you will need to get these products in the door.

Single-vendor

There may be some components of your system architecture wherein it is only practical to source with one vendor. These could include optical storage devices, mainframes, high-security systems and others. The big disadvantage, of course, of sourcing an open system as though it were a proprietary one is that your vendor may have little incentive to behave like an open system vendor. Therefore, this option should be used with care and only when absolutely necessary. With open systems, it is usually better to have at least two suppliers available who can meet your requirements.

Multi-vendor

For most components of your system architecture, you will have a choice of a number of vendors who make products that are both operationally suitable and which exhibit (or at least promise) sufficient open system standards adherence to meet your requirements. More than likely, you will want to avail yourself of this freedom of choice, and also the various protections which market competition provides you. For a given component, therefore, you must decide:

- how many vendors would be an appropriate number to accept, approve or certify?
- how should we conduct such process?

Regarding the number of vendors, you must consider the nature of the component and your present/future requirement for it, as well as the nature of the process you intend to use. Let's structure the consideration of these issues via a few simple questions.

1. *How many of these items will you buy?*
   The more systems you intend to acquire, the more you will want to focus on a vendor's current (and projected future) products, technology base, standards adherence and support capabilities. The greater the value of the total anticipated acquisition, the more important it is that the number of vendors be *optimized.*
2. *What process will we use to decide which vendors we want to deal with*?
   Sometimes it is first easier to decide how to get the vendors in the door before deciding how many to let in. You may decide to use any of the following:

   - Merely *accept* vendors, based on published specifications or their marketing demonstrations
   - *Lab test* wherein you bring in samples of the competing products to your own testing lab for evaluation
   - *Field trial* to try out a product and possibly also other products which go with it
   - *Pilot*, which places the product in a situation as close to actual production as possible

3. *What is the optimum number of vendors for this component*?
   This is a function of two key items:

   - the nature and scale of the product to be procured (it is easier to have ten PC suppliers than ten MRS or mainframe suppliers)
   - the method you intend to use for letting them in (see item 2 above).

In general, the number of vendors should fall with rising product cost and complexity. Also, the more extensive your means of admitting vendor products, the fewer vendors you can handle, all else being equal. Clearly, if you are intending to mount full field trials to certify something even as pedestrian as laser printers, you will be able to manage only a quite finite number of such trials. If, on the other hand, they will just be run through a lab test there might be no problem certifying fifty brands. See Table 2.4 for a list of recommended numbers of vendors.

Table 2.4 Recommended number of trial/pilot sites and vendors

Product	Units to be acquired	Number of vendors
PC	Under 50	2
PC	50–500	3
PC	500–5000	5
PC	Over 5000	10
WS	Under 50	2
WS	50–500	2
WS	500–5000	3
WS	Over 5000	5
MBS	Under 50	1
MBS	50–200	2
MBS	200–500	5
MBS	Over 500	7
MRS	Under 10	1
MRS	10–50	2
MRS	50–200	3
MRS	200–500	5
MRS	Over 500	5
Midframe	1–10	2
Midframe	Over 10	3
Mainframe	Any number	2

Omni vendor

There are some products for which you do not need to go to a great deal of trouble, even if certification is called for. For example, you can easily establish a DOS/Intel-based PC386 or PC486 standard and use paper specifications, quick demonstrations or elementary lab tests to verify compliance with your specifications. Subject only to your plans for technical service, you may find that ten or even twenty suppliers are manageable, particularly if you have a third-party support vendor who has agreed to support anything compliant to your specification.

Software registry concept

One of the most important opportunities presented by an open system strategy is ability to create a central software 'recycling' facility or repository. Where a local line manager asks the local system administrator to find software with a given functionality, that person's first recourse would be to the organization's central software registry. The existence of the registry offers such a workgroup four options:

- REGISTRY: obtain software from the registry at no cost from those packages developed for the same purpose by other workgroups and deposited there.
- MODIFY: obtain software from the registry and customize it to specific local requirements – a copy of each modified package is returned to the registry
- THIRD PARTY: purchase an off-the-shelf package from a third-party software vendor (in this case the registry may wish to negotiate inclusion of a corporate site licence in the deal, so that other workgroups may in future access the application via the above two channels.
- DEVELOP: develop the application from scratch internally or with an external contractor – a copy will flow to the registry.

Naturally, this is only practical if all workgroups have computers running compatible operating systems. It would work if they all had the same proprietary system, but in a large organization it is likely that they will have equipment built by different vendors. Only if all such equipment runs UNIX (or at least a POSIX-compliant or XPG4 branded operating system) would such a 'software commonwealth' be possible.

☛ Because programs for open systems are easy to move between computers, you should set up a central depot for all your computer programs. Then, anyone who needs a new program can look there first. This will save your organization a great deal of money.

☛ The software depot or 'registry' operates not unlike a used car lot. It obtains used software and returns it to circulation. Whether you make users 'buy' the software from the depot or give it to them, this is little more than an internal accounting matter in most cases.

Managing the pace of migration: timing is everything

It is widely implied, but perhaps not well enough stated, in my previous book that the timing of the launch of your open system migration program is critical. To the greatest degree possible, seek a time that will grant you the requisite senior management attention and support. (For example, it would not be wise to commence while they are in the midst of beating off a would-be hostile takeover by a competitor.) It is absolutely crucial to have their support as well as their understanding of how long the process will take. Once you are under way, you will still have to deal with various key pacing factors, as set out below.

Business and external issues

If a hostile takeover (or similar crisis) were to begin *after* you are under way, you might have to slow down the pace of migration or even stop until the situation settles down.

Standards and technology developments

During any point of your journey, standards or technological developments may work to aid or block your path forward. The announcement of a new and attractive (but very proprietary) operating system such as Windows NT may serve to de-focus some of your early supporters. Momentous standards developments, such as the movement of X/Open into the very centre of the UNIX world, may give you cause for pause if you have committed to a particularly arcane or highly vendor-flavoured form of that operating system.

Planning capability and workload

You cannot roll out lab tests, pilots, trials and production systems any faster than you can comprehensively plan them. The first such exercise without a well-grounded plan is almost guaranteed to be your first experience in open systems disaster!

Implementation capacity and workload

Even if your planners are prolific, they can still swamp your implementors on the ground. During any war, the defence department plans far more campaigns than are ever implemented. Some of these plans are needed in case various contingencies arise; others, it would seem, are created just to keep the planners busy – and their pencils sharp – when there is no real war planning to do.

User status and views

You must maintain a pace which is no greater than your user community's ability to continuously assimilate new technology.

Conclusion

Open systems offer the implementor some daunting challenges, but also some wonderful rewards. Early explorers navigated their wooden sailing ships among islands laden with gold, silver, diamonds and many other treasures. However, these islands also tended to have reefs and shoals. Strategic goal-setting, meticulous planning, steering by the stars and careful pacing are the friends of the mariner. They are the friends of the open system implementor too.

3 System functional model: demand and supply

Introduction

This chapter presents a system functional model (SFM) which permits comparison of the system resource requirements of a user (or group of users) with the actual system resources to be provided. The model therefore permits a single configuration, or a group of alternative configurations, to be developed to meet a specific known requirement on the part of a user or user group. These configurations can then be compared in terms of surplus capacity, expandability, performance, cost and other factors.

☛ We will create a model of the different parts of a computer system which will help us to plan, build up and compare systems of various types.

The model addresses the demand for, and supply of, *technology*, *communications* and *support* resources. Regarding resource supply, within each of these three categories a seven-layer stack is hypothesized. Each layer of each stack is dependent upon the one below it. Thus, each system configuration is 'built up' from the lowest levels, in view of the actual requirement the system is to fulfil. The model is designed both for standalone use, as a system configurator, as well as for eventual use in concert with an economic model and an application placement tool.

Objective

The objective is to provide the basis, preliminary definition and initial values (for calibration purposes) for a system component functional model for use in addressing architectural, economic and operational issues in system planning and implementation. The functional model is part of a triad of tools which also includes an economic framework and an application placement framework.

☛ We are trying to create a building-block way of modelling or simulating the relative power of various computers, communications networks and support capabilities. We want to do this so that we can consider what we have in place now and can then consider the likely cost and performance of various alternative future mixes of computers, communications and support. If open systems are genuinely better, the model should allow us to demonstrate that fact.

Requirements and context

The approach taken to develop the SFM was required to take account of the following areas of concern:

- *Supply and demand*: the model must address the capacity or capability of system architecture components in relation to functions and capabilities that users require and actually make use of – in all cases the supply must equal or exceed the demand, otherwise a constraint, reduced level of service or blockage condition exists.

☛ We have to provide enough technology so that the users can do their work effectively and efficiently. If we provide them with too much (beyond a margin for growth) we are wasting money; if we provide them with too little they will have constant capacity problems.

- *Capacity and performance surrogation*: the very large (and rapidly increasing) number of ways to measure the capacity and performance of information technology products makes impractical the creation of a model which directly addresses all such measures. Thus the introduction of new measures and the rapid increase in system rating against existing ones would render the model at best unstable and at worst useless – the model must therefore use a surrogate capacity and performance rating system for technology supplied to the user which is a direct counterpart to an identical system rating actual user requirements – the rating system must be based on units or multiples of some known basket of capabilities provided to a user.

☛ In order to figure out how much to supply, we must first figure out how much is needed. If I measure the capacity of my car's fuel tank in 'quardgets' and you sell fuel only in US gallons, how will I know how much to buy and how will you know how much to sell me? We require a common unit of measure, even if it is imperfect.

- *Utility resources*: some functions and capabilities are provided on an 'outlet' basis in that they are made available, nominally at least, on an equal basis at all times and in all places just as the power supply is provided to all electrical outlets.

☛ Like electrical power, some capabilities are easy to make available everywhere, and to everyone.

- *Local resources*: some functions and capabilities are provided on a 'make at the site' basis in that they are made available solely through the collection, assembly and integration of resources at the site level, be it a workplace, unit, area or enterprise tier site.

☛ Not all computer systems are built top-down; some are built bottom-up.

- *Basic four-tier system architecture*: while two-, three- and four-tier architectures are all valid, the latter is the most challenging and it therefore provides a reasonable basis for development of the SFM.

☛ Tiers are the different sizes of computer system, from small portable and desktop PCs all the way up to very large mainframes. Palmtops, personal digital assistants (e.g. Newton) etc. are not expressly covered in this model, but they could certainly be added where necessary.

- *Tier invisibility*: the user must be required to take minimal cognizance of the tier or tiers on which provided capacity resides or from which it is provided – in the most preferred scenario, processing tier (and communications among them) would be invisible to the user.

☛ Eventually, users will care less about where the computer power comes from (whether from the PC on their own desktop or from a mainframe thousands of miles away) than they do about how they use it. However, for now and for the foreseeable future, use of more distant and lofty computer power still brings with it the associated penalties, particularly where access to close personal assistance is required by novice users.

- *Open computing systems/OSI*: the SFM must recognize the portability, interoperability and scalability offered by open computing systems and the interoperability offered by the OSI model.

☛ There is no point in making a model to help plan open systems if it cannot tell an open system from a proprietary one. You wouldn't hire a guide to take you tiger hunting unless you believed that the guide could tell a tiger from a bear.

- *Support hierarchy*: the SFM must recognize that a hierarchy exists for the provision of user support and that each level of such support depends upon the presence of the levels below it.

☛ Supporting computers is very complicated and demands people with many levels of skills and experience. Generally, they tend to depend on each other, as they each handle only a part of the problem. Our model must cover these people and the professional relationships and dependencies which develop among them.

- *User profile*: when user requirements are reduced to the most basic level, users require three generic sets of capabilities, functions and/or services – these are *technology*, *communications* (connectivity) and *support*.

☛ Users want computing and communications (to connect people and computers in various places together), and they want *other* people to support the computers and communications; they don't want to support it themselves.

- *Tier neutrality*: the SFM must not be biased in favour of, or against, any tier of processing – it is recognized that the comparison among tiers, and the selection of the appropriate tier(s) for a given application is the province of the application placement framework and *not* of the SFM.

☛ Our model should not bias us in favour of large computers, mid-sized ones or small ones. It should help us to treat all of them equally when it comes to planning. Open systems establishes a sort of democracy within the computer world.

- *Vendor neutrality*: the SFM must not either implicitly or explicitly favour or disfavour any vendor, proprietary (or for all intents and purposes proprietary) standard, protocol, means of operation or product.

☛ With so many vendors able to supply open systems products, it is foolish to favour (or become too dependent upon) one vendor.

- *Component interrelationship*: a given component (A) can have three types of relationships with other components:

  - X is the foundation for A (A requires X)
  - A is the foundation for Y (Y requires A)
  - Z is the peer of A (A and Z may interoperate)

☛ In a system architecture-driven environment, the various components have clearly defined relationships and interdependencies. These must be reflected in our model.

- *Capital/operating costs*: any component can be characterized, at a given point in time, as to its capital and operating costs, such that at any site, or for the entire organization, the aggregation of costs for all SFM components represents the total cost of supplying the system resources modelled – these costs do not include incidental or overhead costs and are treated on a capital/operation and maintenance (O&M) basis, not an imputed rent basis.

☛ Even if we do not care, our accountants will care how much these wonderful systems cost.

SFM approach and overview

Key hypotheses

The approach taken in developing the SFM was based on the hypotheses as set out below.

1. It is reasonable to divide the supply of, and demand for, resources associated with information technology (and excluding the data/information itself which is also a resource) into TECHNOLOGY, COMMUNICATIONS and SUPPORT elements. There is thus a demand for – and a supply of – capacity/performance under each of these three categories (Fig. 3.1).

☛ People need the services of computer hardware/software, of communications networks and of those who support them. In turn, it is necessary to arrange how much of each to supply to a given user, workgroup, department or corporation.

2. There is reasonable *parallelism* in the hierarchy of standalone system architecture components (extending from hardware up to user application), the hierarchy of communications (the OSI seven-layer model) and the hierarchy of support components. Dividing the technology and the support elements into the same (somewhat arbitrary) number of levels into which the communications element was earlier divided provides two benefits:

   - there is a reasonable (although not perfect) correspondence across elements at any given level – they are at the very least analogously related, and in most cases the relationship is stronger
   - in all three elements, each level can be shown to be dependent upon the level(s) below it (Fig. 3.2).

	TECHNOLOGY	COMMUNICATIONS	SUPPORT
REQUIRED CAPABILITY	TR	CR	SR
SUPPLIED CAPABILITY	TS	CS	SS

Figure 3.1 Demand and supply model – basic concept.

REQUIRED

TECHNOLOGY (TR)	COMMUNICATIONS (CR)	SUPPORT (SR)

SUPPLIED

TECHNOLOGY (TS)			COMMUNICATIONS (CS)	SUPPORT (SS)
PACKAGE	APPLICATION	APPLICATION	APPLICATION	SELF
	PACKAGE		PRESENTATION	SPECIALIST
		LANGUAGE	SESSION	LANA/MRSA
OPERATING SYSTEM			TRANSPORT	USER-ASSIST
			NETWORK	VENDOR
HARDWARE			DATA LINK	CBI
			PHYSICAL	DOCUMENTATION

Figure 3.2 Demand and supply model.
LANA = LAN Administration
MRSA = Midrange System Administration
CBI = Computer-based Instruction

☛ For each of the computer, communications and support, there are basic resources that you have to have in place before you can provide the more advanced capabilities. The more advanced components depend on the basic ones. This is just like the relationship between the foundation, walls and roof of a house.

3. If a given level of capacity is required of the top level of the seven-layer stack within any of the technology, communications or support stacks then the user (who is supported) must have a 'look-down-shoot-down' ability to 'see' a minimum of that level of capacity at Level 7 and at all levels below it in that stack. For example, note that in Fig. 3.3 Level 5 of the communications stack will not support those above it. (The meaning of values is explained later, but note that this level is a 'bottleneck' not permitting the (CR = 1000) communications demand to be supported all the way down through the stack.) Figure 3.4 demonstrates various alternative mixes of components.

☛ You can not build a fifth storey onto a four storey building unless the underlying structure can support the extra weight. Each storey must be supported by all those below it; each storey must support all those above it. The same type of hierarchy exists in the computer world.

4. For each stack the minimum common level of capability (the minimum supported upwards from Level 1 by *all seven layers*) must equal or exceed the demanded capability at Level 8. Thus, the demanded TECHNOLOGY, COMMUNICATIONS (connectivity) and (user) SUPPORT capabilities can be viewed as a 'Level 8' which is placed above, and is supported by, the previously discussed seven levels. The demand could represent that posed by a single user, a contiguous workgroup (which might employ an MRS), a corporately distributed workgroup or any other collective of user demand.

☛ Rather than build two parallel sets of 'stacks' of demand and supply, we are treating demand as an 'eighth layer' which we superimpose on top of the seven layers of technology, communications and support capability. This is reasonable because we put the seven layers there to support the user in the first place.

User requirements

A set of *general assumptions* is made in this section about basic desktop end-user requirements as follows:

- The least demanding users will wish to run only an application written in a conventional programming language (such as one written in COBOL, FORTRAN or C) – these users are hypothesized to require no less than the capacity of a DOS/Intel standard 80286-based PC, which has been the desktop standard in many organizations for a

REQUIRED

TECHNOLOGY (TR)	COMMUNICATIONS (CR)	SUPPORT (SR)
(5000)	(1000)	(1500)

SUPPLIED

<table>
<tr><th colspan="3">TECHNOLOGY (TS)</th><th>COMMUNICATIONS (CS)</th><th>SUPPORT (SS)</th></tr>
<tr><td rowspan="3">PACKAGE (5000)</td><td>APPLICATION ( )</td><td rowspan="2">APPLICATION ( )</td><td>APPLICATION (1000)</td><td>SELF (1500)</td></tr>
<tr><td rowspan="2">PACKAGE ( )</td><td>PRESENTATION (2000)</td><td>SPECIALIST (1500)</td></tr>
<tr><td>LANGUAGE ( )</td><td>SESSION (500)</td><td>LANA/MRSA (N/A)</td></tr>
<tr><td colspan="3" rowspan="2">OPERATING SYSTEM (5000)</td><td>TRANSPORT (2000)</td><td>USER-ASSIST (4000)</td></tr>
<tr><td>NETWORK (2000)</td><td>VENDOR (2000)</td></tr>
<tr><td colspan="3" rowspan="2">HARDWARE (17000)</td><td>DATA LINK (2000)</td><td>CBI (1500)</td></tr>
<tr><td>PHYSICAL (2000)</td><td>DOCUMENTATION (2000)</td></tr>
</table>

Figure 3.3 Example model values.

The session layer of the communications stack does not provide the required level of support.

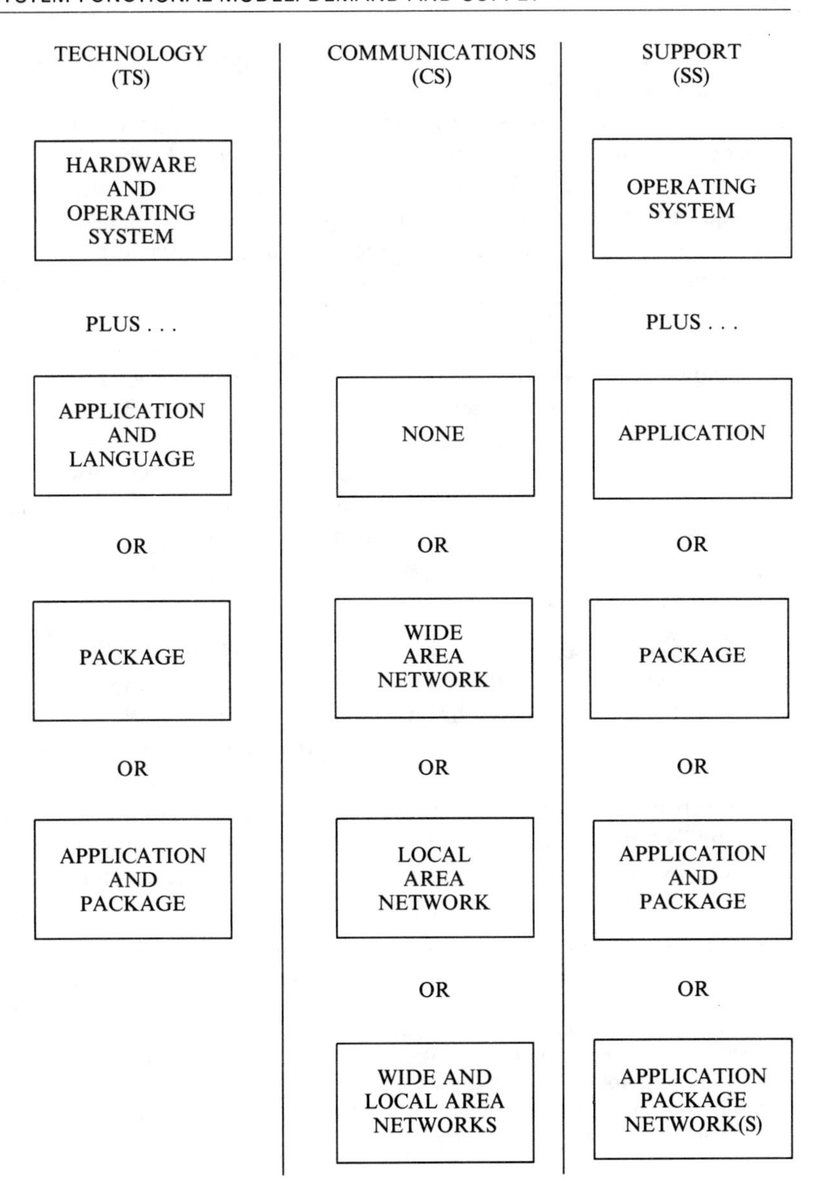

Figure 3.4 Technology supply – alternative components.

period of time (although more powerful systems are now being adopted in large numbers).
- More demanding users will wish to run a package which can serve the user without an application being mounted (such as Lotus 1-2-3 or an office automation (OA) package) – these users are hypothesized to require a more powerful system, where they remain standalone users
- The most demanding users will seek to run an application which is dependent upon a package (such as one written for dBASE or even for standalone ORACLE or a similar RDBMS) – these users have an even higher system resource requirement, where they remain standalone users (throughout much of this book ORACLE is used as a surrogate for whichever RDBMS your organization may have selected for its corporate standard).
- The second class of user above places approximately twice the demands on his or her system than does the first class – the third class places about three times the demands of the first class.

☛ Some users need more computing power than others; this fact can be represented in our model.

It is at once recognized that this 1-2-3 relationship (which is characterized below as a 1000-2000-3000 relationship) is almost certainly an inexact measure of system demands made by the three classes. However, it is believed to be a reasonable 'surrogate' because the ordering of system demand levels (among user classes) is correct. It is also true that the number of highly context-specific variables impacting a given individual's true system resource demands is too large to support a user taxonomy with a small enough number of user classes to support use within the model.

☛ In other words, it is necessary to generalize about the level of demand various users place upon their systems so as to capture the differences in such demands within the model. We have arbitrarily defined the most lowly standalone computer which we think anyone in our organization might productively utilize as a '1', or more precisely, as a '1000'. This provides us a unit of measure, much like the gallon, the horsepower or the mile. The only problem, of course, is that our unit of measure is not really unidimensional but is actually a 'vector sum' of three different kinds of resources: technology, communications and support.

Technology

Referring to Fig. 3.2, it will be noted that hardware comprises the first two tiers, and here represents CPU, memory and storage capabilities primarily, although basic printer capabilities are dealt with as an auxiliary capability. Consideration was given to limiting these items to Level 1 and making Level 2 a linkage layer (I/O devices primarily);

however, such additional complexity brought little benefit in increasing the sensitivity of the model for analytic, simulation or expert system purposes. Clearly, each user must automatically have I/O capability anyway, and this can be safely assumed. In cases of workgroups with PC and MRS equipment (where end-users may access by terminal emulation, OS-to-OS rapport or via vertically integrated applications) this approach could result in double-counting of some or all I/O capabilities.

Levels 3 and 4 represent the operating system (OS) which is, of course, dependent upon the presence and correct functioning of the hardware levels.

☛ Without the hardware, the OS would have to remain on the disks or tape on which it is supplied and would do nobody any good.

Above the OS are three alternative (and in many multi-user systems, co-existing) paths for delivery of technology through to Level 8:

- *Level 5–7 package* such as a spreadsheet, a word processing package or even complete office automation (OA), and in any event a package which does *not* require the user to have a separate application to perform useful work.
- *Level 5–6 package* such as an RDBMS which is combined with a Level 7 application which is either written expressly for the package and/or written in a manner (or with a language/tool) designed for, and thus dependent upon, that package.
- *Level 5 high-level language*, in which is written a Level 6–7 application – such an application is usually more complex (all else being equal) than one created for a package while the language offers less capability than the package to the end-user and thus spans one level and not two or three.

Here, a slight variation on the 'look-down-shoot-down' aspect of the model is required. The hardware and OS levels must support the total weight of capacity provided by summing all Level 5, 6 and 7 capabilities which are placed atop the OS. *Thus, in a multi-user system it is necessary to sum the total of the provided (and actually used) capabilities for all of the three paths described above.* Note, however, that from the user perspective he or she is only deemed to look down through one path. Providing there is no capacity deficiency (against user requirement) in that path, the particular user is deemed to be fully served.

☛ If a hotel chain built three ten-storey hotels, one might have a parking garage built over its tenth storey, another might have a swimming pool and a third could just have a penthouse. All three, however, might be of the same basic design. Of course, the one with a five-level parking garage or a pool with thousands of gallons of water on top would need far more structural reinforcement than the one with a one-storey penthouse. The same principle applies here; the more demanding the higher level applications, the more robust the lower layers must be. The company can either build all three with enough strength to support the heaviest load, or optimize the structures for each of the overhead loads. Overbuilding can be expensive; under-building is flat-out dangerous, especially in earthquake country!

Communications

Here, communications refers to the provision of service to the end-user to a degree required by his or her requirements. Within the OSI model, each layer except Level 1 is dependent upon the layer(s) which are below it. Most open, partially open and closed (proprietary) communications networks render communications services which can be categorized and 'slotted in' according to the OSI model. Of course, there is one important caveat. Few compatible suites or combinations of products fully implement the entire OSI model at this time; however, the model provides a reference and planning guide. Further, the current pace of OSI-capable product releases promises the availability of full suites of OSI products in the mid-1990s.

Communications capability (i.e. the communications services provided by a LAN or WAN) is not viewed as a surrogate, competitor or replacement for a tier of processing, such as MRS. Further, the connectivity services delivered by communications networks and the information processing capacity provided by technology-based systems are not viewed as directly interchangeable. For example, the term 'network processing' as a verb or noun is therefore in most cases not allowable; it is processors which process, not networks. Networks merely link processors (and users) with each other. While they may have internal processing to support communications, networks *themselves* do not process information on behalf of end-users in most circumstances.

☛ Communications hierarchies or stacks are more difficult to understand intuitively, since they are actually 'nested' rather than actually being piled on top of one another.

Support

The provision of support to the user is the most difficult hierarchy to visualize because, unlike the other two, it is built from a foundation of information rather than a foundation of hardware. As we move up the other two stacks we move from hardware to the eventual delivery of capabilities or services to the end-user. Climbing the support stack we move along a continuum from availability of information to theoretical understanding to actual experience on the part of the end-user. Basic information is put to progressively more powerful and dynamic use – and the user increases his or her experience base – as we move up the stack.

Level 1 represents information contained in user and reference *documentation.* Without such hard copy and/or on-line information it would not be possible for the end-user to be supported at all because no one would know how to operate the system, package or application, nor would they know how to access any network.

Level 2 is the much livelier (indeed self-presenting) information found in *computer-based instruction* (CBI), interactive courseware format. It cannot exist without the base documentation first existing. Virtually all present day-systems, packages and applications are at least somewhat self-instructing with dedicated courseware, novice modes, dummy examples etc. Communications networks are considerably less friendly at this time but these too are developing self-instructional or help facilities where necessary.

Level 3 is the *vendor* support of each of the technology and communications elements provided to a given user or user group. In most cases each such vendor will maintain a telephone hot-line, data-line dial-up or other access mechanism permitting the end-user, system administrator and/or the supporting IT professional to obtain supplier assistance. Such supplier assistance may involve walk-through assistance over the telephone, but will almost invariably also involve reference to hard copy or on-line documentation and possibly to CBI courseware as well.

Level 4 is the *user-assist* end-user support facility operated by IT which can be accessed via various methods, but which cannot support the user without the existence not only of a vendor (to whom IT's own unsolvable problems can be referred) but also of documentation to which the user can refer. Again, the existence of CBI material is not absolutely mandatory now, but in the future it may well be seen as such.

Level 5 is the provision (where applicable) to the end-user of the services of a system administrator who provides assistance and support related to hardware, OS, package(s) and application(s). It is only those services that the system administrator provides to the end-user that are counted as part of provided capacity in the SFM. Services provided to the local line manager in such areas as locating new applications or identifying the need for information management planning (or other professional services) are not counted here. They are referred to here as LAN System Administration (LANA) or Midrange System Administration (MRSA).

Level 6 represents a site *specialist* (in the hardware, OS, package and/or application) who may be a part-time vendor representative who visits the site at intervals, a part-time application-specific administrator or other specialist available on an as-needed basis to the end-user. Note that this too is a 'where applicable' tier, since many standalone users and some using LAN-linked PCs and/or multi-user systems dedicated to relatively simple applications may not require such support.

Level 7 represents the minimum degree of *self-sufficiency* training required by the end-user to be able to initiate a system session, invoke an application if necessary, link application and data, perform useful work, obtain further information from or about the application, back up data, restore data, leave the application and terminate the system session. Clearly, the level of support provided by lower levels must meet or exceed the requirements of the user in order for such user to have this degree of self-sufficiency.

Note that the definition of user support specifically excludes hardware and software support to the technology and communications elements, which is considered to be included within (and costed with) those elements.

Supply and demand

Basic unit of service

Within the basic four-tier architecture adopted for our model the lowest or most basic level is the workplace tier. At that tier the most basic service which can be provided to a standalone user is the provision of an intelligent workstation running an OS and an internally developed (or other) application written in a high-level language. Within the model we use a PC286 (PC-compatible with Intel 80286 CPU) baseline, as the lowest level of standalone technology to be provided to any user. Of course, this ignores the case of non-intelligent terminals or X-terminals, but they require a larger system. Of course, many organizations are moving to higher standards. There are nonetheless many users even today for whom the PC286 provides more than adequate capacity and performance. A composite basic workstation has been defined which includes the following significant components:

- 80286 CPU (10–15 MHz)
- Memory (1–2 Mbyte)
- Storage (40 Mbyte)

This is considered to be a baseline system. Note that the system does *not* include LAN or WAN connectivity; nor does it include directly accessible printing (the PC has no at-desk printer and is not connected to one). Included as *access mechanisms* are a colour or monochrome monitor of

EGA/VGA performance level, a graphics board and driver software appropriate to the application, a full 101-key keyboard (bilingual if desired) and a graphical user interface (GUI) such as MS Windows (MSW) although it runs more satisfactorily on a PC386. Also included is a shell or front-end for the operating system for basic file manipulation and to support backup/restore activities by the end-user. Basic CBI for DOS is also included.

Note that the PC286 itself, no matter how widely used, is taken here as an arbitrary baseline. In order to run WordPerfect 5.1 and MSW (and a few other common applications) *adequately*, it will likely be necessary to specify a workstation with at least 2–4 Mbyte of memory. Note, however, that this represents the second (not the lowest) class of demand set out above; such a system would have a score of TS = 2000. For an organization with only a few workers who will have desktop PCs for use on a dedicated basis (one application only, with no 'package') then use of the PC286 as the base is reasonable. However if a 4 Mbyte system is viewed as the minimum *for any user* then it may be desirable to change the baseline and treat the above-defined PC286 as a special 'under-class' of system. It should also be pointed out that whatever baseline is chosen, a Macintosh with approximately equivalent power could be considered a reasonable substitute; nothing in the model confines it to the DOS/Intel architecture at the low end. In order to begin a system of measurement and modelling, it is necessary to start somewhere.

This PC286 configuration is sufficient to permit the user to mount an operating system (almost certainly a single image of MS-DOS in this case) and an internally created or third-party dedicated application program written in a high-level language. Note that this program is not a 'package', such as Lotus, WordPerfect etc. However, if the program is non-trivial in its processing and data requirements a significant percentage of the CPU, memory and storage capacity of the machine will be utilized, subject of course to the addressing restrictions of the DOS operating system.

Of course, this degree of capability can be provided to the user by means other than providing the above-cited equipment on a desktop basis. For example, the user instead could be provided with either of the following:

- A graphics-capable terminal, star-wired to the UNIX operating system and the same application, running in multi-user mode on an MRS, where all of the CPU, memory and storage capability is provided by the MRS as this user's share of its resources.
- A graphics-capable diskless workstation client, star-wired to an MRS running UNIX and acting as a server in which case part of the

capability would come from the MRS and part from the workstation, where the workstation itself might need, for example, only half the capacity of the PC286 if the other half was provided by the server.
- A graphics-capable diskless workstation client, star-wired to a micro-based server (MBS) running OS/2 or NT and acting as a server in which case part of the capability would come from the MBS and part from the workstation, where the workstation itself might need, for example, only half the capacity of the PC286 if the other half was provided by the server.

For SFM purposes, all of the above are held to be *precise equivalents* to the PC286 in terms of level of service for the user who requires only a single custom application and the ability to manipulate, back up and restore files. Access to the application running under UNIX on a midframe or mainframe is a *near equivalent*, but it does require X.25 or more advanced OSI network access with its inherent support requirement, which is discussed below. Similarly, access to the application, under MVS, on an ALS or ELS also requires additional resources, here IBM 37XX, 3174 and 3278 access, whether real or emulated.

While the PC286 obviously provides the user some surplus capacity and the ability to progress later to other basic standalone applications, its most significant provision of utility to the user is in the three areas of CPU, memory and storage. Without any one of these being at or close to the levels cited above, our hypothetical application will not run adequately or even may not run at all. The PC286 is therefore declared to provide a basic unit of user service, which is set at 1000 'Technology Supply' or TS points.

This rating of TS = 1000 could be broken down as follows:

- CPU TS = 400
- Memory TS = 200
- Storage TS = 400

There can be considerable debate, even if the concept of a baseline of 1000 points is accepted, as to the relative contributions of the CPU, memory and storage components of this basic level of service. Any division is at least somewhat arbitrary because it assigns the same type of points both to apples and to oranges. It also excludes the input and output mechanisms, without which the user could not access these system resources. Further, changes in the price/performance within and among categories over time may distort this point allocation. Nonetheless, several important factors should be noted:

- Whatever measure of technological capacity and performance is adopted what is crucial is that it be *consistently extended and applied* – if a user has a three-times-greater requirement than 1000 and is provided a workstation with a TS = 3000 rating then we would expect the measures for each CPU, memory and storage to be exactly (or at least approximately) three times those of the PC286 (indeed this is the easiest way of initially calibrating the model), not only against the power of various systems but against the workload imposed in various application scenarios.
- The *lowest common denominator* or basic unit of service, of which no fractions exist (where provision of a complete service to a user is concerned but not as relates to part-services such as printers), should:
  - Be set at 1 or some ten-multiple of 1 (1000 was chosen to give adequate scope for full definition of part-services on an individual and workgroup level)
  - Be the arena in which changes, over time, in the relative price and performance of its CPU, memory and storage components are resolved either by reallocating the basic point score among these components or by actually increasing the total above the unitary (here 1000) level – it does not matter which strategy is chosen providing it is consistently applied, however increasing the value of the unitary score allows impending deficiencies in the installed equipment base, particularly at the MRS tier and above, to be rapidly highlighted since new demand points can be compared directly to old supply points
  - Be adjusted by comparing the CPU, memory and storage requirements needed to run with the current GUI to those required to run with a future (more resource-consumptive) GUI at the time the latter GUI is introduced (for example, the new GUI might boost the basic user requirement from 1000 to 1200)

What must be considered is the practical capacity of a system, which addresses not how much capacity exists in the system but how much is actually made available and used by the user.

☛ Regarding practical capacity, an example drawn from another field may be of assistance. If a locomotive has a diesel engine rated at 3000 h.p. but the aggregate capacity of the electric traction motors driven by its generator is only 2000 h.p., then no more than the equivalent of 2000 h.p. of tractive force can reach the rails. The traction motors serve as the weakest link in the chain and therefore represent a capacity/performance bottleneck in the engine/generator/traction motor/wheel hierarchy or stack.

Similarly, it is clear that a PC286 is functioning in a sub-optimal mode when running DOS, which cannot make use of all of the Intel 80286's capacity. Nonetheless, the system must (and does) provide sufficient capacity/performance to run the non-trivial custom application we have hypothesized. A well-equipped PC386 can run either DOS or UNIX. On this machine DOS addresses an even smaller percentage of the total capacity of the system. In the latter case of the PC386, DOS is the bottleneck in the system hierarchy, which is defined below.

Beyond basic service

As was implied above, many (indeed most) users will not have a requirement which is purely standalone. Even those users who have what is an equivalent (work-alone) requirement, but who do not have a PC or a direct equivalent (local connection to server or MRS), will require SNA, X.25 or other network connection to a larger midframe or mainframe host. Of course, many users will have specific LAN or WAN requirements over and above the basic requirement of being connected to whatever system serves as the primary host.

☛ Where you have to use a network to get even your basic (1000 level computing capability) we do *not* count that network as being anything other than a part of a 'virtual PC' supplied to your desktop.

A user who needs to run one custom application, and who therefore requires the PC286 or either of the above-cited direct equivalents, has a 'technology requirement' score of TR = 1000. The standalone user has a 'communications requirement' of CR = 0 and a basic 'support requirement' of SR = 1000. Thus, this end-user's User Demand Profile (UDP) can be expressed in TR/CR/SR format as 1000/0/1000. Note that TR, CR and SR points are *not* interchangeable since each represents a different type of requirement.

☛ Since we believe that networks are for communicating, not for processing, we cannot 'trade-off' among computers communications and support any more than you can swap back and forth among petrol, motor oil and windshield washer antifreeze for your car.

Modes of supply

There are three modes of supply contemplated within the model: the ABSOLUTE, RELATIVE and REPLICATIVE modes. These are loosely (but not absolutely) correlated with the technology, communications and support elements of the model.

The ABSOLUTE mode relates to the capacity and performance capability that derives directly from locally or centrally installed hardware and software. For example, if there are three users whose total requirement can be expressed as 3000/0/3000 then installation of three PC286 systems would fulfil the technology element of this requirement. If a fourth user arrives on the scene, the only way to meet his or her requirements is to acquire *another* system. No system, no capacity. No capacity, no service. This would also apply to LAN equipment/software which has certain base overheads and is then configured on a per-user basis. A LAN configured for 20 users will *not* serve 21. An MRS with a 64 user UNIX license will *not* serve the 65th logon.

☛ From the local end-users' (or their managers') perspective this translates into the dictum: 'If you want it, you will have to buy it!'

The RELATIVE mode relates primarily to corporate and network-oriented services such as the communications services provided by the X.25, SNA and future OSI networks, as well as the user support services made available by vendors, USER-Assist and site specialists who can – except in extreme cases – provide their responsive capabilities as easily to 50 as to 20 users.

☛ These are 'tappable' resources which 'extend relative to the demand' in the same way that 110 V or 220 V electricity is available from various outlets throughout an office. More electric power for Person A does not mean less for Person B. Of course, an electric utility and an organization's USER-Assist service (within the IT Group) share the requirement to establish capacity to meet peak demand with a reasonable level of service, and in this regard the total or extended demand is an important planning number. However it is only the end-user whose access position is $N+1$ who does not receive the prescribed level of service (perhaps he gets no service at all), where N is system capacity. The supplied resource can be held to have previously 'expanded' to accommodate each new user until the level of N users was reached. There are various economic schools of thought regarding pricing of utility type services. From the local perspective this mode is governed by the dictum: 'It is already there; you get your required capacity just by tapping into it!'

The REPLICATIVE mode is basically a trivial mode. This relates most closely to multi-user systems. If an MRS is upgraded from 16 to 48 users then for each of the 32 new potential users all of the required documentation must be made available. If, according to contractual arrangements, the documentation may be copied, there is no capital cost to upgrade this support component because not only is photocopying an O&M expense, but it is outside the system costing realm anyway. (On the other hand, if it must be purchased it is ABSOLUTE mode, since each copy of the documentation must be paid for and failure to acquire causes a supply deficiency.) Full CBI support for the new users is required, but in upping the user limits on the OS, they will automatically gain access to the CBI packages(s) and are thereby served. Similarly, with the issue of licences for the larger number of users, the vendor(s) may automatically extend support to the new users. In such case, the supply mode is REPLICATIVE. If the vendor charges for each new user, the supply mode is ABSOLUTE. REPLICATIVE differs from RELATIVE in that there are no central or corporate capacity constraints; if the vendor has given permission to copy a manual, it is likely that an almost unlimited number of copies may be made for internal use within the organization.

☛ Some aspects of support can be easily replicated.

It may also be possible to hypothesize a MIXED mode. Consider an enterprise level system (ELS) (a.k.a. mainframe) which, from the end-user perspective, offers almost unlimited capacity. However, each user increases the load on the system and (linearly or S-curve-wise) the continual addition of users, with their various degrees of imposed workload, moves the system through various degrees of service degradation and towards the crash state. Normally whichever is the most imposing of the various I/O, CPU, OS or similar limitations results in such a crash. In this sense the system behaves in RELATIVE mode. However, in some areas, such as user logons or time-share monitoring/control or even specific applications there may be absolute limits impacting the system. Perhaps one of these system elements will only tolerate 3022 users and the 3023rd user will simply not gain access, but will not 'crash' the system for the existing full load of users. In such a case, the system element is behaving in ABSOLUTE mode. Generally, a complex system or facility, such as an MRS or a larger system, will represent the MIXED mode when all of its components are considered as a composite.

☛ No matter how much capacity is supplied there is always the potential for that one additional user who becomes the proverbial 'straw that breaks the camel's back'.

User taxonomy

Individual users

Using the above model of the standalone user's requirement and that person's supplied basket of PC286 resources as a basis, it is possible to vary each of the technology, communications and support requirements to address more sophisticated requirements. A summary of the twelve generic user types (A–L) is displayed in Table 3.1. Each of these generic user types represents the requirement of a single standalone or networked user as well as the local (and share of other) resources applied to fulfil such requirement.

USER A is served by a PC286 with a 10–15 Mhz CPU, 1–2 MB memory and 40 MB hard disk. This user requires only the language and application elements of Layers 5–7 of the technology stack and therefore bypasses the others. No communications capability is provided. The

Table 3.1 User requirement classes – summary

CLASS	TECHNOLOGY	COMMUNICATIONS	SUPPORT
	HARDWARE AND OPERATING SYSTEM PLUS . . .		OPERATING SYSTEM PLUS . . .
A	APPLICATION (1000)	NONE (0)	APPLICATION (1000)
B	APPLICATION (1000)	LAN (1000)	APPLICATION/LAN (2000)
C	APPLICATION (1000)	WAN (500)	APPLICATION/WAN (1500)
D	APPLICATION (1000)	LAN/WAN (1500)	APPLICATION/LAN/WAN (2500)
E	PACKAGE (2000)	NONE (0)	PACKAGE (1500)
F	PACKAGE (2000)	LAN (1000)	PACKAGE/LAN (2500)
G	PACKAGE (2000)	WAN (500)	PACKAGE/WAN (2000)
H	PACKAGE (2000)	LAN/WAN (1500)	PACKAGE/LAN/WAN (3000)
I	APPLICATION/PACKAGE (3000)	NONE (0)	APPLICATION/PACKAGE (2500)
J	APPLICATION/PACKAGE (3000)	LAN (1000)	APPLICATION/PACKAGE/LAN (3500)
K	APPLICATION/PACKAGE (3000)	WAN (500)	APPLICATION/PACKAGE/WAN (3000)
L	APPLICATION/PACKAGE (3000)	LAN/WAN (1500)	APPLICATION/PACKAGE/LAN/WAN (4000)

documentation includes an application user manual and DOS user and reference manuals. A DOS CBI package is provided. Vendor support for DOS and the application is available by telephone to the user, although first off-site recourse would normally be to USER-Assist, which is shown at its full capacity to support the most demanding single user, not just at the level required for this user. Note that throughout these single-user examples the system administrator level of support is not applicable except that some support exists where a LAN is provided. A visiting specialist provides initial training and is available on-call, probably on a more accessible basis than the original equipment/product vendors. Often this will be the VAR, ISV or SI firm which provided the products to the site. The user receives the self-sufficiency training described above, for the application and for DOS (Fig. 3.5).

☛ This is the most basic user. All other users have higher requirements for computing, communications and support in one or more respects.

USER B is identical to USER A except that LAN connection of the PC is required. The LAN capability is held to be 50% implemented on the PC (LAN card and software) and 50% implemented outside the PC (LAN server/controller, LAN OS, wiring and connectors). However, here the total capacity provided to the end-user is not broken down according to its origin. A later LAN server example clarifies this. At the software levels each of the PC LAN software and central (server/controller) LAN software contribute half of the CS = 1000 requirement. Similarly, at Level 1, each of the PC and the server/controller contribute one half of the same value. Note that the support requirements have doubled and a LAN administrator is available on-site at all times. Documentation for the user now includes LAN user and reference manuals, vendors include the LAN vendor and the user is trained in LAN access and basic use. Basic LAN CBI support is provided on the PC. See Fig. 3.6 and Table 3.1.

In Table 3.1, USER C does not have LAN access but is connected to a WAN, for which user documentation and at least basic CBI support is provided both for DOS and the WAN. The communications and support requirements are lower than for the LAN but higher than for standalone operation. Here, there is no local administrator. Note that while the basic user requirement (CR value) for WAN connection will stay constant at 500/user the supply values – at different tiers – for different types of WAN (e.g. SNA, X.25, future OSI network) can be varied. This becomes significant in multi-user situations.

USER D is basically a combination of the USER B and USER C profiles, which again holds processing requirements constant but combines the requirements for LAN and WAN communications and support services.

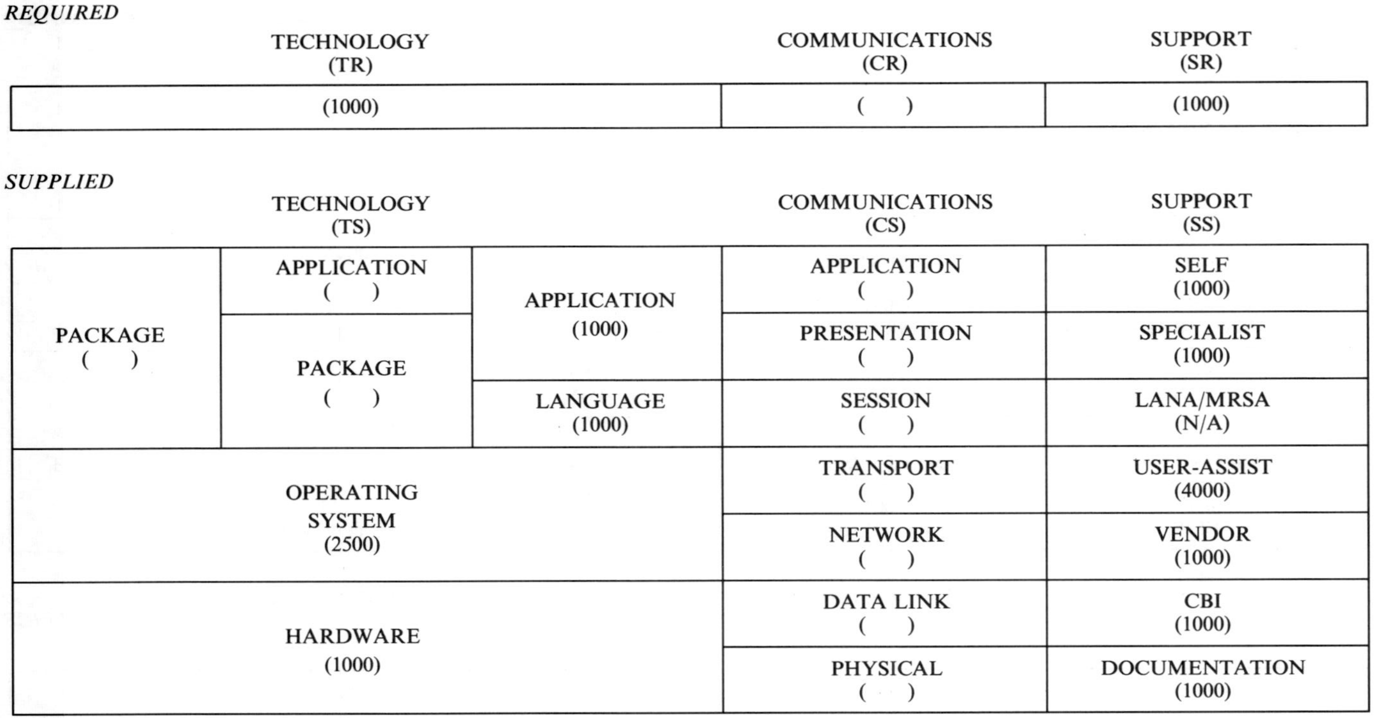

REQUIRED

TECHNOLOGY (TR)	COMMUNICATIONS (CR)	SUPPORT (SR)
(1000)	( )	(1000)

SUPPLIED

<table>
<tr><th colspan="3">TECHNOLOGY (TS)</th><th>COMMUNICATIONS (CS)</th><th>SUPPORT (SS)</th></tr>
<tr><td rowspan="3">PACKAGE
( )</td><td>APPLICATION
( )</td><td rowspan="2">APPLICATION
(1000)</td><td>APPLICATION
( )</td><td>SELF
(1000)</td></tr>
<tr><td rowspan="2">PACKAGE
( )</td><td>PRESENTATION
( )</td><td>SPECIALIST
(1000)</td></tr>
<tr><td>LANGUAGE
(1000)</td><td>SESSION
( )</td><td>LANA/MRSA
(N/A)</td></tr>
<tr><td colspan="3" rowspan="2">OPERATING
SYSTEM
(2500)</td><td>TRANSPORT
( )</td><td>USER-ASSIST
(4000)</td></tr>
<tr><td>NETWORK
( )</td><td>VENDOR
(1000)</td></tr>
<tr><td colspan="3" rowspan="2">HARDWARE
(1000)</td><td>DATA LINK
( )</td><td>CBI
(1000)</td></tr>
<tr><td>PHYSICAL
( )</td><td>DOCUMENTATION
(1000)</td></tr>
</table>

Figure 3.5 User Class A.

REQUIRED

TECHNOLOGY (TR)	COMMUNICATIONS (CR)	SUPPORT (SR)
(1000)	1000)	(2000)

SUPPLIED

<table>
<tr><th colspan="3">TECHNOLOGY (TS)</th><th>COMMUNICATIONS (CS)</th><th>SUPPORT (SS)</th></tr>
<tr><td rowspan="3">PACKAGE
( )</td><td>APPLICATION
( )</td><td rowspan="2">APPLICATION
(1000)</td><td>APPLICATION
(1000)</td><td>SELF
(2000)</td></tr>
<tr><td rowspan="2">PACKAGE
( )</td><td>PRESENTATION
(1000)</td><td>SPECIALIST
(2000)</td></tr>
<tr><td>LANGUAGE
(1000)</td><td>SESSION
(1000)</td><td>LANA/MRSA
(2000)</td></tr>
<tr><td colspan="3" rowspan="2">OPERATING
SYSTEM
(2500)</td><td>TRANSPORT
(1000)</td><td>USER-ASSIST
(4000)</td></tr>
<tr><td>NETWORK
(1000)</td><td>VENDOR
(2000)</td></tr>
<tr><td colspan="3" rowspan="2">HARDWARE
(1000)</td><td>DATA LINK
(1000)</td><td>CBI
(2000)</td></tr>
<tr><td>PHYSICAL
(1000)</td><td>DOCUMENTATION
(2000)</td></tr>
</table>

Figure 3.6 User Class B.

USERS E–H are identical to USERS A–D except that here they each require the services of a package which is much more sophisticated than a single, dedicated application program. The requirements of such a package alone dictate the provision of a more powerful system. The supplied PC386 has a TS = 2000 rating with the following breakdown:

- 80386-based CPU (20–30 MHz) TS = 700
- Memory (3–4 MB) TS = 400
- Storage (80 MB) TS = 800
- Maths coprocessor TS = 100

Clearly, Lotus 1-2-3 or WordPerfect can be run on a lesser machine. However, recall that we have assumed a full GUI, full OS bilingualism and a CBI as a part of the user's standard access to applications wherever they reside and a portion of the resources provided within each system are dedicated to this feature. It is also intended to provide the user with the same level of service (including response time) as the PC286 provided to the less demanding users. Experience in many organizations has shown that users of packages such as Lotus, WordPerfect or single-user desktop manager packages also generate greater demands for support, even without LAN or WAN connection, than do users of fully customized and dedicated applications. Here and throughout the user types discussed the minimum support requirement is generally based on an upward extension to the higher levels of the support stack of the value of Level 1 – Documentation, which is derived from the appropriate combination of the following:

- Application (User) SS = 500
- Package (User) SS = 1000
- DOS (User/Ref) SS = 500
- UNIX (User/Ref) SS = 1000
- LAN (User/Ref) SS = 1000

The values are similar for CBI except that for DOS CBI SS = 1000 and CBI is not available for applications. DOS and UNIX CBI packages are shown at equal values because known examples of these cover approximately the same material for each operating system, whereas UNIX documentation is much more profuse than DOS documentation in most cases.

For the OS/2 operating system, a TS rating for a single-user system would exceed that of DOS (2500), but would be considerably less than the rating for UNIX (5000). Based upon a comparison of task capability and flexibility a reasonable preliminary estimate would be approximately 3500 for a single-user workstation. Note that where OS/2 is used on a LAN server, the supported OA or RDBMS package too is acting in the server mode. In such a mode (unlike the multi-user mode) the package's total capability is *not* a full linear extrapolation of the capacity requirement of each user. If 50% of the package/application work is performed on each of the server and PC then only one half of the total capacity is required to be server-resident. Further, because OS/2 is not multi-user and is more restricted in its multi-tasking than is UNIX, a linear extrapolation of OS/2 capacity rating for the number of users may overstate the true OS/2 capability. Based on previous benchmarking or service comparisons, or based on published benchmarks or other published data, it will be necessary to establish a multiplier discount for OS/2 as follows:

$$\text{TS(OS/SVR)} = \text{TS(OS/PC)} \times \text{CLIENTS} \times 0.\text{NN}$$

where

TS(OS/SVR) = TS rating of server OS/2 operating system
TS(OS/PC) = TS rating of PC OS/2 operating system
CLIENTS = maximum number of clients server can accommodate
0.NN = OS/2 multiplier discount factor

☛ Within our model, we can show that the OS/2 operating system is more powerful than DOS but considerably less powerful than UNIX.

Similarly, for Windows NT (the user interface which ate the operating system), its capability would exceed that of OS/2 but, for the foreseeable future at least, is expected to be far below UNIX. Thus, NT could be 'placed' within our model without undue difficulty, once its capabilities (particularly in respect of its server role) become more widely understood and properly measured. This has yet to occur.

A review of Table 3.1 will indicate that support requirements are rising faster than technology or communications requirements as we consider progressively more demanding user requirement scenarios. This is certainly true to the real life situation.

☛ Linear increases in user demands and in combinations of computing and communications actually supplied tend to bring more than linear increases in total support requirements. Too frequently, gloriously conceived and designed systems fail in the field for lack of adequate user support.

USERS I–L are identical to the previous two sets of users except that here the requirement is to run an application program written for (and dependent upon) a package. Their technology requirements profile is based on TR = 3000. For purposes of comparison, assume a single-user program written for ORACLE. Note that in certain cases, postulated CBI capabilities may be inadequate if no CBI is provided for in-house or third-party applications.

User groups

Because the UDP indicates the maximum load that a user can be expected to place on technology, communications and support resources, for a given user the category chosen from those set out above must always be his or her most demanding usage mode. For example, if someone uses a word processing package in standalone mode but also uses an RDBMS application while connected to a LAN, then they are classed as a USER J, not USER E. Similarly, when aggregating users for a multi-user application and/or system it is necessary to state each user's UDP at its conceivable maximum. The aggregation of all individual user requirements should therefore represent the peak or maximum of system utilization which can be expected. Thus, the total TR/CR/SR point score for all users should overstate the actual system requirement, thereby providing the users (or at least those with the least demanding requirements) some growth capabilities.

☛ We can add together the individual requirements of a group of users to get their group requirements.

This concept is not new. The normal means of sizing an MRS would be to assess the number of simultaneous ORACLE users, the number of OA users and the number of other users and then address specific capacity in terms of CPU, memory, storage, I/O etc. However, using three point scores as a surrogate for this raw capacity requirement permits implicit trade-offs among these factors to be considered, for example, in comparing a multi-user versus a client-server RDBMS paradigm, particularly where UNIX is used as a common operating system in both cases. It is known that for a given application a client-server approach will require greater *total* (MRS and PC based) storage than in multi-user mode. Similarly, a client-server implementation of a given RDBMS application may require more *total* (MRS and PC based) processing resources.

Fig. 3.7 demonstrates the aggregation of individual user requirements for specification of a 50 user MRS. In this example there will be 25 users of Class J and 10 of Class F. Their aggregate UDP is 95 000/35 000/100 000. It will be noted that the per-user requirement for functionality of the application and RDBMS is 3000. The application itself can be extended by REPLICATIVE mode; once it is multi-user it can have any number of users limited only by hardware, OS and RDBMS capacity. The RDBMS is governed by the ABSOLUTE mode; it must provide sufficient capacity for *each* user who will use it simultaneously. We assume here that ORACLE is rated only for the number of users required (25) or else is restricted by the system administrator to this number of

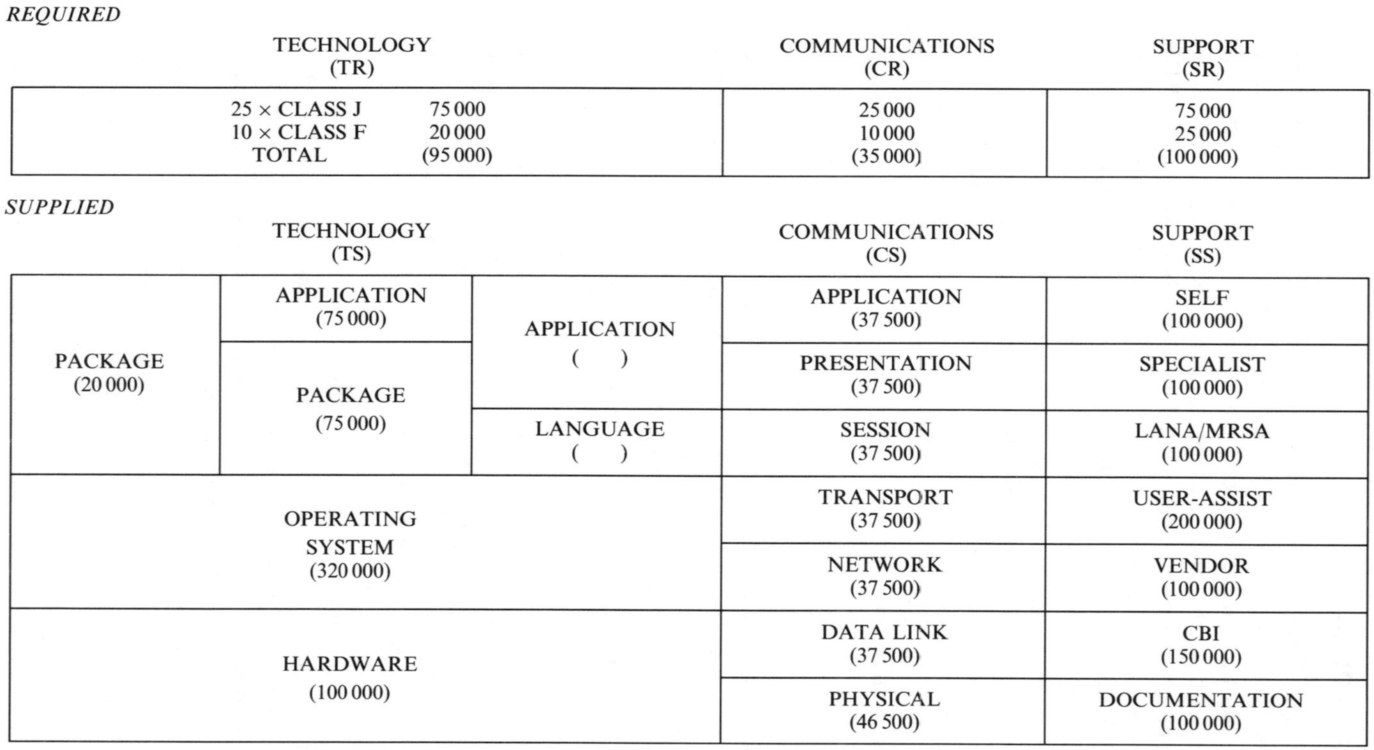

REQUIRED

<table>
<tr><th>TECHNOLOGY
(TR)</th><th>COMMUNICATIONS
(CR)</th><th>SUPPORT
(SR)</th></tr>
<tr><td>25 × CLASS J 75 000
10 × CLASS F 20 000
TOTAL (95 000)</td><td>25 000
10 000
(35 000)</td><td>75 000
25 000
(100 000)</td></tr>
</table>

SUPPLIED

<table>
<tr><th colspan="3">TECHNOLOGY
(TS)</th><th>COMMUNICATIONS
(CS)</th><th>SUPPORT
(SS)</th></tr>
<tr><td rowspan="3">PACKAGE
(20 000)</td><td>APPLICATION
(75 000)</td><td rowspan="2">APPLICATION
( )</td><td>APPLICATION
(37 500)</td><td>SELF
(100 000)</td></tr>
<tr><td rowspan="2">PACKAGE
(75 000)</td><td>PRESENTATION
(37 500)</td><td>SPECIALIST
(100 000)</td></tr>
<tr><td>LANGUAGE
( )</td><td>SESSION
(37 500)</td><td>LANA/MRSA
(100 000)</td></tr>
<tr><td colspan="3" rowspan="2">OPERATING
SYSTEM
(320 000)</td><td>TRANSPORT
(37 500)</td><td>USER-ASSIST
(200 000)</td></tr>
<tr><td>NETWORK
(37 500)</td><td>VENDOR
(100 000)</td></tr>
<tr><td colspan="3" rowspan="2">HARDWARE
(100 000)</td><td>DATA LINK
(37 500)</td><td>CBI
(150 000)</td></tr>
<tr><td>PHYSICAL
(46 500)</td><td>DOCUMENTATION
(100 000)</td></tr>
</table>

Figure 3.7 Midrange system (50 users).

simultaneous logons; otherwise the RDBMS could certainly crash the system. Note that each of the application and RDBMS provides the users with TS = 75 000 while the parallel path of the package (OA) provides another TS = 20 000. All lower levels of the stack must, however, bear the combined total TS = 95 000 load imposed by all paths through Levels 5–7. Table 3.2 provides a generalized framework for making such calculations.

Table 3.2 Site aggregate demand profile calculation format

USER GROUP USER TYPE	NUMBER	TR	CR	SR
A	5	15 000	5 000	7 500
B				
C				
D				
E				
F				
G				
H				
I				
J				
K				
L				
OVERHEAD ITEMS TYPE OF OVERHEAD ITEM		TR	CR	SR
SYSTEM ADMIN RESOURCE		5 000	3 000	10 000
OTHER SYSTEM OVERHEAD				
PRINT RESOURCE				
ADDITIONAL NETWORK ACCESS				
OTHER				
TOTAL REQUIREMENTS		20 000	8 000	17 500

While the basic specifications for MRS equipment might contemplate that some 50-user systems will have only 10 or 15 simultaneous ORACLE users, the machine indicated here is clearly a heavy duty system. Reference to Class I–L users (where UNIX is provided in single-user mode) will indicate that the utility provided to the single user is not only double that of DOS but also exceeds the immediate requirements of even the most demanding (application/RDBMS) user. This is an accurate statement of the situation because future insertion of a more powerful GUI or other user assets which increased the basic hardware requirement for these classes would in all likelihood *not* require an upgrade to basic

OS capacity. When UNIX is extended to multiple users we use the ABSOLUTE mode of extension, in so far as licensing is concerned, and we ignore the fact that even most basic 'single user' UNIX will in most cases carry the potential to host up to eight users. In our example, the 64-user rated capacity of the operating system provides a TS = 320 000 score (i.e. 64 × TS = 5 000, as discussed above). Thus, the OS is in no way a bottleneck within the stack; on the contrary, it is correctly shown as having the highest 'TS throughput' capacity of the entire technology stack.

Regarding hardware, the internal breakdown of the TS = 100 000 capacity provided by the MRS is as follows:

- CPU (25–30 MHz) (RISC) TS = 25 000
- Auxiliary CPU (same) TS = 25 000
- Memory (32 Mbyte) TS = 3 500
- Auxiliary memory (same) TS = 3 500
- Storage (1000 Mbyte) TS = 10 000
- External storage (2000 Mbyte) TS = 20 000
- Printer laser) (2) TS = 6 000
- Printer (pin) (2) TS = 2 000
- FP processor (50 user capacity) TS = 5 000

With respect to communications, the required CS = 35 000 LAN capabilities are provided as follows:

- PC LAN software (35 × CS = 500) CS = 17 500
- PC LAN cards (35 × CS = 500) CS = 17 500
- Server software (for 40 users) CS = 20 000
- LAN controller card (40 user) CS = 20 000
- LAN connector panel CS = 4 000
- LAN boosters CS = 4 000
- LAN connectors CS = 1 000

Thus, the total capability provided at Level 1 (hardware) of the communications stack is CS = 46 500, while at Levels 2–7 (software) it is CS = 37 500.

Note that at this site a system administrator is available on a half-time basis (supporting both the MRS and LAN) and this provides a rating of SS = 100 000. Normal ABSOLUTE, RELATIVE and REPLICATIVE extensions apply to the various other categories, as discussed above. Consider another example, this one for a 20 user, LAN-equipped MRS with 10 RDBMS/application users (Class J) and four OA users (Class F) expected to be the maximum simultaneous load. Here, total UDP is 38 000/14 000/40 000. It will be noted that this system offers considerably more surplus capacity (relative to its package and package/application loading) than does the previous example. Here, UNIX is licensed for 32 users and a different 'gauge' of hardware is employed, specifically:

- CPU (25–30 MHz) (RISC) TS = 10 000
- Auxiliary CPU (same) TS = 10 000
- Memory (16 Mbyte) TS = 1 750
- Aux Memory (16 Mbyte) TS = 1 750
- Storage (800 Mbyte) TS = 8 000
- External Storage (1000 Mbyte) TS = 10 000
- Printer (Laser) (2) TS = 6 000
- Printer (Pin) (1) TS = 1 000
- FP coprocessor (20 user capacity) TS = 2 000

This MRS is rated at TS = 50 500, approximately one half the capability of the larger system discussed above.

The LAN breakdown, to provide the required CS = 14 000 is as set out below:

- PC LAN software (14 × CS = 500) CS = 7 000
- PC LAN cards (14 × CS = 500) CS = 7 000
- Server software (for 20 users) CS = 10 000
- LAN controller card (20 user) CS = 10 000
- LAN connector panel CS = 4 000
- LAN boosters CS = 4 000
- LAN connectors CS = 1 000

In this example, CS = 26 000 capability is provided at Level 1 of the stack and above that level the provided capability is CS = 17 000. The system administrator is available for approximately 20% of the time of a full-time person, which can be expressed as 0.20 PY.

Finally, Fig. 3.8 profiles a requirement for 30 Class J users who require LAN access however who will employ the client-server (versus the true multi-user) model of RDBMS operation. Note that in this example we are profiling *only the server* requirements. Therefore, instead of the expected 90 000/30 000/90 000 demand on the shared system we see only a demand of 45 000/15 000/45 000. This, of course, assumes that each of the 30 users possesses a workstation capable of providing each of them with the other half of the required capacity. Further, they are assumed to have the PC-based LAN card and LAN access software already. (Note that if we were profiling *total* site requirements for these users, under the client-server model, we would indeed count the PC resources as part of each stack.)

In this case we have chosen to use a 'de-rated' MRS-based system with a single CPU in the server role rather than a high-end PC engine. UNIX

REQUIRED

TECHNOLOGY (TR)	COMMUNICATIONS (CR)	SUPPORT (SR)
30 × CLASS J 90 000 LESS: PCs −45 000 TOTAL (45 000)	30 000 −15 000 (15 000)	90 000 −45 000 (45 000)

SUPPLIED

<table>
<tr><th colspan="3">TECHNOLOGY (TS)</th><th>COMMUNICATIONS (CS)</th><th>SUPPORT (SS)</th></tr>
<tr><td rowspan="3">PACKAGE
(8000)</td><td>APPLICATION
(45 000)</td><td rowspan="2">APPLICATION
( )</td><td>APPLICATION
(15 000)</td><td>SELF
(45 000)</td></tr>
<tr><td rowspan="2">PACKAGE
(45 000)</td><td>PRESENTATION
(15 000)</td><td>SPECIALIST
(45 000)</td></tr>
<tr><td>LANGUAGE
( )</td><td>SESSION
(15 000)</td><td>LANA/MRSA
(45 000)</td></tr>
<tr><td colspan="3" rowspan="2">OPERATING
SYSTEM
(160 000)</td><td>TRANSPORT
(15 000)</td><td>USER-ASSIST
(120 000)</td></tr>
<tr><td>NETWORK
(15 000)</td><td>VENDOR
(90 000)</td></tr>
<tr><td colspan="3" rowspan="2">HARDWARE
(49 200)</td><td>DATA LINK
(15 000)</td><td>CBI
(45 000)</td></tr>
<tr><td>PHYSICAL
(24 000)</td><td>DOCUMENTATION
(45 000)</td></tr>
</table>

Figure 3.8 Dedicated server (for 30 clients).

is here rated as in the previous example. The supplied TS = 49 200 of hardware capability is broken out as follows:

- CPU (25–30 MHz) (RISC) TS = 15 000
- Memory (12 Mbyte) TS = 1 200
- Storage (600 Mbyte) TS = 6 000
- External Storage (2000 Mbyte) TS = 20 000
- Printer (Laser) (2) TS = 6 000
- Printer (Pin) (1) TS = 1 000

The LAN breakdown, to provide the required MRS portion of LAN capability, which is rated at CS = 15 000, is as set out below:

- Server software (for 30 users) CS = 15 000
- LAN controller card (30 user) CS = 15 000
- LAN connector panel CS = 4 000
- LAN boosters CS = 4 000
- LAN connectors CS = 1 000

Actual capacity provided at Level 1 is CS = 24 000 and at Levels 2–7 is CS = 15 000. Larger area and enterprise level systems are shown in the next section.

Technology component profiles

Preliminary definitions, using the SFM, for various components of a typical system architecture are found in Figs. 3.9–3.11. As indicated above, in order to build up a multi-part architecture component (such as an MRS) it is necessary to make a composite overlay of such items as the basic system, RDBMS package, OA package and others. Therefore some of the components – as presented – do not stand alone. The 'basic information' section of each profile provides information about the component's generic descriptions as well as its place in the architectural hierarchy with respect to what supports it, what it supports and its peers. The 'Cost information' section details capital and O&M costs for the component. The costs set out in these examples are primarily for demonstration purposes although they are believed to be reasonably close to actual costs. Note that the wide area networks were not costed. Training costs for end-users were calculated on the basis of a $150 direct cost for each DOS user and a $300 direct cost per user for each of UNIX, applications and packages. The 'Key attribute/characteristic value summary' section breaks down the component into its various parts in terms of TS, CS and SS supply point scores. These are then mapped into the triple stack structure in the subsequent 'Component stack profiles' section.

Basic Information

GENERIC NAME: PERSONAL COMPUTER ACRONYM: PC286

VENDOR: DESCRIPTION:

SUPPORTS: LANG, PKG, APPL

PEER1: WS CLASS: SYSTEM PEER2: PC386

REQUIRES:

SUPPLY MODE: ABS

Cost Information		
CAPITAL		
Corporate Planning/Acquisition/Management	$	500
Purchase/Delivery		3 000
Installation/System Integration/Commissioning		200
Orientation/Training		150
TOTAL CAPITAL	$	3 850

OPERATION AND MAINTENANCE		
Corporate Support	$	100
Annual Licence Fees		
Lease/Rental		
Software Support/Upgrade		
Hardware Maintenance		300
Administrator Salary and Related O&M		
Adminstration or Facility Management Contract		
Supplies and Expendables		100
Re-location/Re-installation		50
Electrical		20
Dedicated facility or Support Equipment		
Other ( )		
Other ( )		
TOTAL O&M	$	570

Key Attribute/Characteristic Value Summary

ATTRIBUTE	VALUE	TS	CS	SS
CPU	80286/10-15 MHz	400		
MEMORY	1-2 MB	200		
STORAGE	40 MB	400		
OS	DOS 3.3	2 500		
DOCUMENTATION	USER/REF			500
CBI	DOS			1 000
VENDOR SUPPORT	DOS			500
TRAINING	DOS			500

Component stack profiles

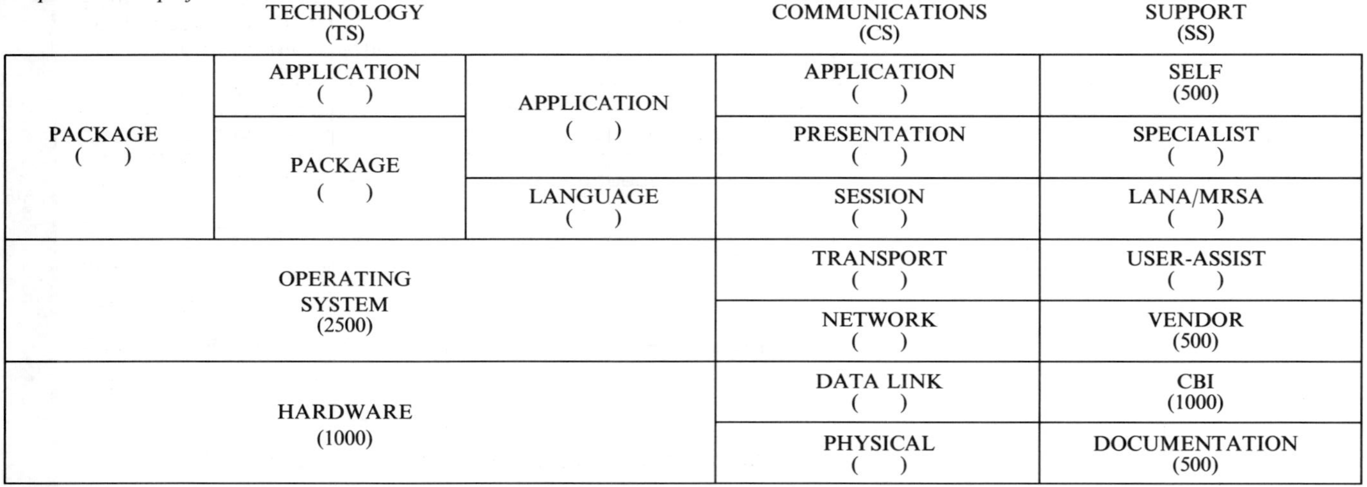

<table>
<tr><th colspan="3">TECHNOLOGY (TS)</th><th>COMMUNICATIONS (CS)</th><th>SUPPORT (SS)</th></tr>
<tr><td rowspan="3">PACKAGE ( )</td><td>APPLICATION ( )</td><td rowspan="2">APPLICATION ( )</td><td>APPLICATION ( )</td><td>SELF (500)</td></tr>
<tr><td rowspan="2">PACKAGE ( )</td><td>PRESENTATION ( )</td><td>SPECIALIST ( )</td></tr>
<tr><td>LANGUAGE ( )</td><td>SESSION ( )</td><td>LANA/MRSA ( )</td></tr>
<tr><td colspan="3" rowspan="2">OPERATING SYSTEM (2500)</td><td>TRANSPORT ( )</td><td>USER-ASSIST ( )</td></tr>
<tr><td>NETWORK ( )</td><td>VENDOR (500)</td></tr>
<tr><td colspan="3" rowspan="2">HARDWARE (1000)</td><td>DATA LINK ( )</td><td>CBI (1000)</td></tr>
<tr><td>PHYSICAL ( )</td><td>DOCUMENTATION (500)</td></tr>
</table>

Figure 3.9 Architecture component profile – PC286.

Basic Information

GENERIC NAME: MIDRANGE SYSTEM (50 USER) ACRONYM: MRS/50

VENDOR: DESCRIPTION:

SUPPORTS: LANG, PKG, APPL

PEER1: MRS/20 CLASS: SYSTEM PEER2: MRS/30

REQUIRES:

SUPPLY MODE: MIXED

Cost Information	
CAPITAL	
Corporate Planning/Acquisition/Management	$ 20 000
Purchase/Delivery	140 000
Installation/System Integration/Commissioning	15 000
Orientation/Training	25 000
TOTAL CAPITAL	$ 200 000
OPERATION AND MAINTENANCE	
Corporate Support	$ 10 000
Annual Licence Fees	
Lease/Rental	
Software Support/Upgrade	2 000
Hardware Maintenance	
Administrator Salary and Related O&M	40 000
Adminstration or Facility Management Contract	
Supplies and Expendables	5 000
Re-location/Re-installation	2 500
Electrical	500
Dedicated facility or Support Equipment	500
Other ( )	2 000
Other ( )	
TOTAL O&M	$ 62 500

Key Attribute/Characteristic Value Summary

ATTRIBUTE	VALUE	TS	CS	SS
CPU	RISC/25-30 MHz	25 000		
CPU	(SAME)	25 000		
MEMORY	32 MB	3 500		
AUX. MEMORY	32 MB	3 500		
STORAGE	1000 MB	10 000		
EXT. STORAGE	2 000 MB	20 000		
FP PROCESSOR		5 000		
PRINTER (LSR)	2	6 000		
PRINTER (PIN)	2	2 000		
SNA EMUL.	10 USER		5 000	
X.25 EMUL.	30 USER		15 000	
OS	UNIX	320 000		
LANG	C/COBOL	50 000		
DOCUMENTATION	USER/REF/2 WAN			100 000
CBI	UNIX WAN			100 000
MRSA				100 000
VENDOR SUPT.	UNIX WAN			100 000
TRAINING	UNIX WAN			100 000

Figure 3.10 Architecture component profile – midrange system.

Basic Information

GENERIC NAME: DATABASE SYSTEM FOR MRS (50 USER) ACRONYM: RDBMS

VENDOR: DESCRIPTION: RDBMS ENGINE V 6.1 PRODUCT SET

SUPPORTS: APPLICATION

PEER1: OA PKG CLASS: PACKAGE PEER2: LANG/APPL

REQUIRES: SYSTEM

SUPPLY MODE: MIXED

Cost Information

CAPITAL

Corporate Planning/Acquisition/Management	$	
Purchase/Delivery		50000
Installation/System Integration/Commissioning		
Orientation/Training.(Admin/User)		20000
TOTAL CAPITAL	$	70000

OPERATION AND MAINTENANCE

Corporate Support	$	
Annual Licence Fees		3000
Lease/Rental		
Software Support/Upgrade		3000
Hardware Maintenance		
Administrator Salary and Related O&M		
Adminstration or Facility Management Contract		
Supplies and Expendables		
Re-location/Re-installation		
Electrical		
Dedicated facility or Support Equipment		
Other ( )		
Other ( )		
TOTAL O&M	$	6000

Key Attribute/Characteristic Value Summary

ATTRIBUTE VALUE	TS	CS	SS
RDBMS ENGINE 50 USER	150000		
DOCUMENTATION USER			50000
CBI USER			50000
VENDOR SUPPORT USER			50000
TRAINING USER			50000

Component stack profiles

<table>
<tr><th colspan="3">TECHNOLOGY (TS)</th><th>COMMUNICATIONS (CS)</th><th>SUPPORT (SS)</th></tr>
<tr><td rowspan="3">PACKAGE ( )</td><td>APPLICATION ( )</td><td rowspan="2">APPLICATION ( )</td><td>APPLICATION ( )</td><td>SELF (50 000)</td></tr>
<tr><td rowspan="2">PACKAGE (150 000)</td><td>PRESENTATION ( )</td><td>SPECIALIST ( )</td></tr>
<tr><td>LANGUAGE ( )</td><td>SESSION ( )</td><td>LANA/MRSA ( )</td></tr>
<tr><td colspan="3" rowspan="2">OPERATING SYSTEM ( )</td><td>TRANSPORT ( )</td><td>USER-ASSIST ( )</td></tr>
<tr><td>NETWORK ( )</td><td>VENDOR (50 000)</td></tr>
<tr><td colspan="3" rowspan="2">HARDWARE ( )</td><td>DATA LINK ( )</td><td>CBI (50 000)</td></tr>
<tr><td>PHYSICAL ( )</td><td>DOCUMENTATION (50 000)</td></tr>
</table>

Figure 3.11 Architecture component profile – RDBMS package/MRS.

Note that what we will call the central IT USER-Assist service was not included here as a specific component (even though it relates to most others), but could be considered a RELATIVE mode resource since total or all-up capability provision does not exceed SS = 4000/user. This also raises the important point of *threshold coverage* of user requirements. An installed MRS with a TS = 90 000 capacity covers the requirements of 30 application/RDBMS users. When some of these users are using the OA package they are, of course, over-covered but as stated earlier, each user must be covered at his or her highest individual requirement. Similarly, USER-Assist provides coverage which meets or exceeds the SR point rating of each individual user's support stack. As with any utility-provided service, capability or capacity, it may be possible to tailor USER-Assist delivery capability only in terms of meeting peak requirements; issues of joint costing and marginal pricing might therefore arise not only in justifying, funding and budgeting a much more thoroughly capable USER-Assist service, but also in addressing the cost/benefit of alternative application placements or system acquisitions.

Implementation guidance

1. It will be seen from the above examples that one advantage of the approach is that current, imminent or potential bottlenecks in terms of technology (hardware, operating system, package and application), communications and support can be detected readily without the requirement to, for example, go back and check again the limitation on the number of users in an application, a LAN licence or similar. This becomes more important when the potential for a semi- or fully-automated version of the model is considered. However, it is necessary to obtain an experience-based understanding of the degree of explicit or implicit trade-off (particularly among the three key hardware resources) that the model should be allowed to exercise.
2. Note that in the development of point score ratings for multi-user systems the TOTAL resource requirement (TR) for the combination of user classes was first considered. Then, required memory and storage were extended linearly in terms of their TS ratings and printers and floating point coprocessors were added where necessary (i.e. for all but pure servers). Then, the TS rating for an appropriate CPU was found as the remainder by deducting thus-far supplied TS from total anticipated TS, which is equal to TR. Then, key performance parameters (MIPS, MFLOPS, AIMS etc.) for the type of CPU found in such a system were compared with those of the unitary system (PC286).

It is strongly recommended that the calibration of model components, within the technology stack, for your organization be performed in this order:

- hardware (PC286, MRS, ALS, ELS)
- operating systems (DOS, OS2, UNIX, VMS, MVS)
- RDBMS engines (ORACLE, SYBASE etc.)
- other packages (such as OA)
- applications written in 3GL/4GL
- applications written for packages (e.g. for ORACLE)

3. When specifying a system for a given user group, it is desirable that the total TS 'throughput' capability (looking down from Level 8 through the technology stack) should exceed actual stated requirements by 30–40%. This is necessary due to the potentials for any one or more of the following:

   - growth in number of users of a package or application
   - growth in number of applications each user uses, particularly in concert with the RDBMS
   - growth in data-set size for RDBMS applications and other applications
   - extension of the definition of the basic level of service (if adding a better GUI next year increases the level of service demanded for a Class A user from 1000 to 1200 then application of the same GUI, if it were available, for UNIX on an MRS (for users accessing it with VT-100 terminals, X/terminals or whatever) could very possibly increase TS resource demands by less than or equal to the same 20% increase, and such a change could come almost instantaneously).

4. While the division of communications resource provision responsibility equally between the PC and server for a LAN is somewhat arbitrary, recent experience in various organizations has shown this to be a reasonable planning estimate. Based on your own experience, it will be possible to state more precisely the demand a given application will place on PC and server processing and communications resources and thus to determine whether a given LAN server and PC configuration can or cannot host a given application written for such an environment.
5. There is no question that the structure of the model lends itself readily to implementation in a simulation-capable (and hopefully AI-capable) database environment. It will be noted, for example, that the *total* or final stacks for a client-server environment represent an overlay or combination of those for a number of individual components

including PCs, MRS or MBS, LAN components, applications, RDBMS package, OA package and system administrator.

6. The use of a colour graphics front-end would be highly desirable. Such a package could allow alternative configurations to be proposed for a given load combination and could (using green, gold and red graphics) indicate surplus capacity, pending and actual bottlenecks at various locations in each of three stacks for each candidate configuration. Even more importantly, such a package could allow the architectural pros and cons of differing approaches to be demonstrated readily to the user.

Conclusion

This chapter has presented a means of 'baselining' a nominal or minimum level of user demand placed upon computing, communications and support resources, and a way of measuring what is actually supplied against such demand under each of these three categories. The model can be used for single-user, client-server and multi-user service models and for all four tiers of equipment. Nothing confines the model to any particular computer architecture or operating system. Note, however, that finalization of the model design, determination of the actual baseline to be used and full calibration of the model are all beyond the scope of this book. The user needs to optimize his or her model customization efforts to the expected beneficial outcome of using the model.

4 Economic model of information technology

Introduction

This chapter sets out a number of economic issues related to application placement and system acquisition within the user organization and where possible proposes a resolution or approach in response to each. Each issue is followed by a brief commentary headed with the (☛) symbol.

In building an economic case for a particular acquisition of an open system acquisition (be it application and/or machine), there are several tools at the reader's disposal:

- The menu of costs, opportunities and benefits cited in Chapter 2 for use at the corporate level, many of which can also be applied to a specific case.
- The economic element of the SFM in Chapter 3, which will hopefully impose some order and discipline in the costing of dissimilar alternatives which are equally potent in terms of TS/CS/SS capacity.
- The guidance provided in this chapter, which addresses a selection of the more 'prickly' issues which will be encountered in constructing the business case.

Business case/cost justification

The business case/cost justification approach now prevalent for project justification within many public and private sector organizations today does not address wider benefits to a given project, where these are beyond the scope of the anticipated benefits of whatever induced the project in the first place.

☛ In the case of an MRS which is installed to run a given application, but is also sized so as to be able to provide OA service to all users, the cost of such a system is usually higher than for one able to run the application *only*. Therefore, *ceteris paribus*, the first system would have a lower payback, but this would underestimate the total project benefit to the unit unless OA was expressly considered. Therefore, use of the business case method alone is not appropriate for analysis of an application placement and/or system acquisition. It will not address important ancillary (or tertiary) benefits of the investment and it may in some cases be too narrowly focused on the specifics of the business at hand, without taking sufficient cognizance of the potentials for improvements to other business, or even conducting new business.

Sunk cost

When an investment was made previously it is considered to be a 'sunk cost' because it cannot be cleanly extricated and turned to another use.

☛ It is reasonable to use the sunk cost concept in our approach but *not* at the outset of an investment. Using the above example, if we are going to go beyond pure business case analysis in assessing whether or not to make an investment, we cannot treat the surplus capacity of an MRS, *vis-à-vis* a given application, (which thus allows it to run OA) as a sunk cost on Day 1 of production and thus say that the OA capability was 'free' to the workgroup.

Opportunity cost

Conventional economics normally recognizes an opportunity cost when a given investment is made or decision is taken, where this precludes other courses of action.

☛ To the degree practical, this should be accomplished in the framework being set out within this economic model.

Effects of contention

The effects of contention or delay which are imposed on each user of a system because that system is at or near its rated capacity must be considered in evaluating two candidate systems, one expected and one not expected to have such a problem.

☛ The effects of contention can be divided into three distinct types, in economic terms:

- The costs of time delays while users are idle, while they take longer than otherwise required to process a given work item (WI) or a group of WIs or the cost of lost productivity when the processing of further work items is simply precluded.
- The cost of user frustration and consequent abandonment of the system which may scuttle many of the projected benefits from an application and/or from a general use package (such as OA).
- The planning and disruption costs of implementing an expansion, upgrade or replacement of the system *within* the project's planned life and *before* it would

otherwise have been undertaken. (It is necessary to assess the risk – in probability terms – of, for example, an overload situation generating such a scenario when a micro-based server (MBS) is provided instead of an MRS.)

Changes in workload

From a more technical perspective, there is the issue of assessing the vulnerability of a given configuration such as MBS/PC or MRS/PC (and the analysis relevant to it) to a change in a factor such as the degree of workload division between the workgroup system and the individual workplace systems.

☛ This aspect of the vulnerability or robustness of the analysis itself can only be based on technical experience, both within your organization and elsewhere. However, the risk of an overload situation – where it can be reasonably estimated or surrogated – *must* be included in the calculation of the potential costs of a candidate system for a project with an inherent risk of an overload situation. To omit it, when another candidate system is provably less susceptible to the same risk, is to overstate the potential benefit of the first candidate system. Thus, the difference in the risk rates of such an overload problem might be considered – in a probability table – against the three types of disruption costs considered above.

Cost of a security breach

The potential cost of a security breach, and the inherently higher total operational cost (where security has a provable value) of a lower security system (where a given level of security is desired or required) must be considered.

☛ It is necessary to look at a possible security breach in one or both of the following two ways:

- The probability of various degrees of security and associated costs.
- The 0/1 possibility of an absolute breach, treated as a threshold.

In practice, most security analysis begins with a systematic and dispassionate assessment of the threats against which it is desired to protect the system, along with its data, applications and users.

Inter-system trade-offs in processing tier

In considering inter-system trade-offs of processing power, as in a client-server/PC environment or multi-user/PC environment both time and tier elements must be considered.

☛ The degree of contiguity (closeness) of two systems in terms of TIME (i.e. when they are planned, installed and upgraded) and in terms of TIER (according to a three- or four-tier system architecture) both impact the degree of vertical (inter-tier) power trade-off possible when planning an application placement or system acquisition. Full expansion of this concept is not always possible at the time of analysis because it includes adding a temporal element to the three-axis portability model cited in Chapter 2.

Benefits of improved configuration flexibility

A key issue in assessing the benefits of open systems is how to show the benefits of improved configuration flexibility as one progresses from PCs through the various systems up to a mainframe. As one moves 'up-tier' there is progressive lessening of physical, connectivity and power restrictions to expansion or upgrade of a system. Traded off against such improvements, of course, are the greater gross costs of providing the progressively larger systems.

☛ A simple approach would be to consider the cost and probability of different generic types of upgrades against different original system baselines (i.e. starting from within different tiers), to perform a (hopefully normalizable) sensitivity analysis which would not be tier-origin specific. This is easier said than done. In general, larger systems offer fewer bars to expansion/upgrade and they thus reduce or eliminate the over-capacitation costs discussed earlier. The issue is to determine what portion of the marginal cost of a larger system (in whatever terms it is measured) should be attributed to the increased configurational flexibility attribute of a larger versus a smaller system (e.g. MRS over WS or PC).

Benefits of workgroup versus individual systems

It is necessary to consider how to measure the benefits of workgroup versus individual system environments. These benefits include workspace sharing, the potential for co-editing and the potential for actual cooperative work (co-working).

☛ The existing literature does not suggest a coherent and consistent means of accomplishing this. This chapter suggests a possible approach, based on work item (WI) analysis. Some readers may be able to substitute a better system based on their own experience and requirements.

Joint costing

There is also the issue of joint costing of capabilities provided to the individual user where these represent a combination of a dedicated system and an allocated share of a larger system.

☛ Refer to Fig. 4.1. Here, we consider two users. USER 1 is provided desktop capability at the TS = 1000 level, but also shares in a higher tier multi-user system which also provides TS = 1000 to that user. While the TS rating system, at the desktop level, does not specifically break down I/O resources, it does assume their existence. As set out in the SFM, once the user has received (for example) a PC286, that person has already been provided with basic 'access' resources, here a keyboard and screen. Thus, when the integrator of the larger system later arrives to provide that same user with TS = 1000 of further (shared) resources, the user is found to be already at least somewhat equipped. The integrator is therefore not really forced to dedicate (for example) a further new keyboard and screen to this user. In fact, it would be quite counter-productive to do so if this meant the user now had to live with duplicated interfaces on the desktop. The user is, in one sense, provided with a full measure of TS = 1000 × 2 here, although the implied basic access resources do not need to be duplicated. If an estimating allowance is normally made for the provision of resources which permit the user to access the shared system, then the cost of providing this TS = 1000 will be somewhat overestimated because of the latent (pre-installed) capacity already on this user's desktop. However, conversely, as pointed out in Chapter 3, not considering this user's I/O issues at all (or even breaking it down as a separate TS component at the desktop level) would also be counter-productive. Indeed, this may give rise to the conundrum that there is no perfect or evidently optimal theoretical approach to solving this problem of 'pre-ordained coincidence of desktop resources' within the previous environment and the as-envisioned state once the new installation is complete. In Fig.4.1, USER 2 is provided with a greater shared resource but, as with USER 1, does not lose the initial PC286.

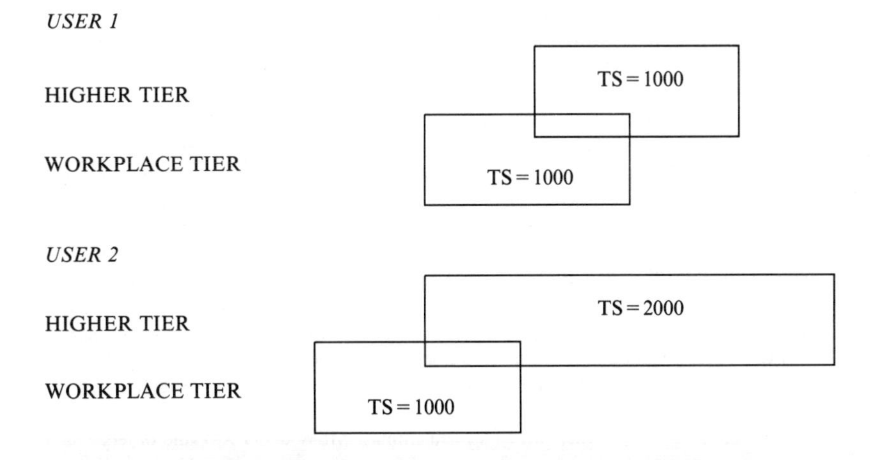

Figure 4.1 Shared resources in the system functional model.

Stuck cost

Figure 4.1 also raises the issue of what we will call 'stuck cost'. The PC286 user, when an MBS or MRS is subsequently installed, will in all likelihood stoutly resist any suggestion that this new installation should herald the removal of the original PC in exchange for anything else, even an X-terminal. Thus, the original investment has become 'stuck' to the user and is very difficult (both operationally and politically) to 'un-stick'. This is just as true even if the new multi-user system will provide all of the functionality the user really needs and the current personal desktop applications are either simply not mission critical for the workgroup and/or they can be uploaded to the MRS and run under DOS emulation within UNIX.

☛ While this is similar to the sunk cost concept, there is a difference. With sunk cost, we are talking about an already-made investment which is seen to contribute to (or detract from) the stream of costs and benefits anticipated from the investment currently under analysis. A stuck cost investment may or may not do so. For example, if a collection of single-user applications on a group of PC286s is replaced by a full multi-user application on a dedicated MRS (which runs no OA and no other applications at all and is intended never to do so) then the PC286s used for the desktop version could be replaced by X-terminals or even non-intelligent (a.k.a. 'dumb') terminals. In such circumstances users might still resist the removal of PCs even if very small or even trivial applications are all that remain on them. Note that this also assumes no evolution of officially sanctioned independent desktop applications or packages and is therefore admittedly a pure or extreme case. It is, however, possible. In such a situation we are now providing too much capacity to each user, of which they will not be required to make near-term use. Should the cost of such unused capacity be somehow offset against the stream of benefits arising from the new system until further productive PC applications are acquired and the PCs are again being productively employed, as PCs and not just intelligent terminals? There would appear to be a theoretical basis for doing so, but the practical justification for doing so may be a little hard to explain to the outraged users whose PCs you now declare you intend to remove! Consider, too, the fact that system users (for any size of system) have their own equivalent of 'Parkinson's Law': the application load will expand to fill the available capacity. Of course, this says nothing about the usefulness or even legitimacy of such applications. They may be computer games or similar applications.

Now consider a more practical example. You want to install five applications, running together on a UNIX MRS, each of which requires the X-Windows (XW) interface, but the users really love their MS Windows (MSW) and do not want to change. Anyway, some of their PC286s and PC386s would have no hope of running XW comfortably without removing almost everything else. They also claim they do not need or want all of that ornate XW functionality. You counter that the new applications actually use some of this extra functionality and it is therefore essential that they have it. Users reply that you have no business bringing them an application that does not use MSW as its front end in the first place.

There are software products which let you emulate MSW under XW and others which promote other types of coexistence. The cost of any such product should be included as a part of your project cost. While the proceeds from transferring MSW licences (even where permitted) or returning them to your supplier may be small, not even these can be counted as a saving due to your project unless you can satisfy the users that they are not losing functionality. Otherwise, you have an unrecoverable stuck cost.

In many cases, you will be forced to address this issue without resolving it satisfactorily.

Cost of a forced upgrade

An even wider issue raised by the same scenario is that of corollary threshold updates. The growth of the PC's share of the workload in an MBS/PC configuration (running principally or only in the client-server service model) could outstrip average or even total PC expansion capability (as limited by chassis expansion slots and/or your budget). This would result in a forced, unplanned and otherwise undesired wholesale upgrade to more powerful desktop systems. In such circumstances, what part of this upgrade cost (after netting out the practically achievable resale value of the existing systems) should be counted as a (risk-weighted probability) monetary cost of the original MBS/PC configuration, when its technical, operational and economic risks were considered at the outset, as compared with those of a more expandable system. The latter does not impose such limitations.

☛ We run into some danger, at this point, of what might be characterized as 'platform bigotry' or bias in favour of systems no smaller than an MRS because, in one sense, any costs accounted against the probability of such an upgrade are indeed the costs of not having selected an MRS-class platform in the first place. These 'risk and danger' costs can be attributed only to standalone, pseudo-LAN, LAN-only (no significant server) and MBS/LAN/PC (client-server) configurations. More operational experience with different types of configuration histories – in cases of units with similar or identical evolution of resource demands over time – is necessary before anything more than a very poor surrogate of this value can be estimated. We do know, however, that we cannot avoid the issue.

In theory, we should be able to capture the economic inefficiency of wholescale logic replication on all PCs in an MBS/PC configuration on a total cost and total TS basis, providing there was not a high latent surplus capacity across the desktops at the time the MBS/PC configuration was first implemented. Certainly, this will be easier where total capacity to be provided to most users is TS = 2000 or TS = 3000, not TS = 1000. This can be carried one step forward to a 0/1 situation at the most extreme case. There is undoubtedly some implicit (and likely high) opportunity cost for a workgroup to deprive itself needlessly of the opportunity for smooth UNIX-based migration from client-server to multi-user where this is required in the future by choosing OS/2 or a similarly restrictive operating system (versus UNIX) now.

When a (small) true multi-user application is forced up-tier by the absolute inability to have the workgroup system (i.e. an MBS) expand to accommodate its logic, user and/or data set growth, there will be incremental costs to expanded demand on larger systems. This too is a part of the economic reality.

Delayed migration anomaly

Juxtaposed across the previous few factors is the delayed migration anomaly and its associated costing issues. Where a unit (expressly so as to leave open the possibility of migrating to future multi-user operation) installs an MRS, instead of an MBS, there initially may be a surplus of capacity even if the MRS is minimally configured. Unless and until this capacity is taken up and used, its associated costs are the cost of keeping open a migration path from client-server to multi-user applications, as well as to larger MRS equipment without starting over. Such costs represent a part of the capital and O&M of the system and should be weighed against the benefits of reduced (or even zero) risk of over-capacitating the system and/or having no easy growth path. Of course, this is an issue which has system, user/operational and economic components. In many cases, the TS element of the SFM will be of assistance here, for if we can readily postulate the workgroup system/PC load sharing, we should be able to identify more easily the surplus TS capacity provided during the early years of the system's life cycle. The ratio of surplus to total TS capacity then renders a cost share which we can allocate – over those early years – to the purpose of buying us 'expandability'.

☛ Once we know how much we have paid for the luxury of being able to migrate without chaos in the future, we need to remember that the capacity may actually be used for one or both of:

- Addressing future capacity shortfalls at the desktop level by changing the split of workload between the workgroup and workplace tiers, but obviously within the confines of the client-server model, the RDBMS (as-implemented) and the actual implementation in question
- Facilitating the migration from client-server to multi-user.

The benefits of the first potential use are obvious: we might forestall or even forego a wholesale desktop upgrade with a clear, readily measurable and significant saving. The costs of the second can be more difficult to fathom, particularly if a change in service model connotes significant changes to the application.

Dedicated and general-purpose platforms

If a functional group at corporate headquarters (such as Human Resources) demands that each line organization have Application X (which runs only under UNIX) and if Workgroup A has no UNIX-based MBS or MRS, then how should the costs of the resulting newly acquired system be allocated? Often, this is also a polite way of asking who should pay for it. The arrival of an MRS may bring with it the need for (and hence the services of) a para-professional or professional system administrator, the ability to run OA and likely at least a moderate capability to run other applications. If Application X requires an RDBMS, then the door is wide open; all one needs is the Value Added Reseller (VAR) catalogue of the RDBMS vendor.

☛ For any of the sub-tiers of the workgroup and workplace tiers, the arrival of a system funded by (or for) one purpose, but which has the capability to be readily put to other uses may require some joint costing. Allocating the entire cost of such a system to one application is seldom the correct approach. The erstwhile dedicated system will soon sprout other applications.

Consider a situation wherein a 50-user MRS is installed to support:

- 30 users each with requirements for Application X, which is rated at TR = 3000, and of course is satisfied by TS = 3000 per user; and
- 20 users who have OA requirements only (TR = 2000/TS = 2000).

All 50 users have an OA requirement. An MRS configured strictly to meet the OA requirements would have a rating of TS = 100 000, not the TS = 130 000 to TS = 150 000 rating which would be necessary to meet the combined requirements of the workgroup, also accommodating Application X. The TS = 130 000 rating is the minimum for a joint-use system; TS = 90 000 for the 30 concurrent application/package users and TS = 40 000 for 20 concurrent OA users. Therefore the minimum *marginal* capacity to extend an OA-only system to provide also for the needs of the 30 application/package users is TS = 30 000 (TS = 130 000 – TS = 100 000) which is TS = 10 000 per user. Conversely, if Application X had first been used to justify the system (i.e. it was first purchased for that purpose and not OA) then a capacity of TS = 90 000 would have been adequate, *but only for the original 30 users*. They would also get access to OA services for no additional resource expenditure.

Now let us expand the system to add 20 more OA users. Suppose that no more than five of the original 30 users require Application X at one time. Therefore their highest concurrent draw on the system is TS = 15 000 (TS = 3000 × 5). The other 45 users would require TS = 90 000 if all logged on concurrently to the OA package. Thus, the incremental cost, in technical terms, of granting 20 more users access to OA services (i.e. to top up the capacity of the system – for everyone to use OA – would be a mere TS = 15 000 or TS = 750 per user. Naturally, a system able to provide full flexibility in user access (offering the original 30 users their choice of OA or Application X at any time) would require an addition of TS = 40 000, the full native requirement of TS = 2000 for each of the incremental users thus added, for a total of TS = 130 000. Here, we have simplified somewhat by considering only hardware and operating system resources. It is of course true that SFM Levels 5–7 investment for the OA package itself is an incremental cost in the case where the system is first purchased only to run Application X. Conversely, expanding an OA-only machine also connotes acquisition of both Application X and its underlying RDBMS.

It may be possible to establish a general theorem of marginal capability provision and costing as set out below.

1. Every information worker has the right to be provided with a basic desktop capability of TS = 1000 together with the support which this inflects. This is provided first and does not require justification.

2. Every information worker who does not function primarily or exclusively in a work-alone mode has the right to be provided basic LAN and/or WAN connectivity of (between CS = 500 and CS = 2000 together with the inflected support) as required to facilitate intra- and extra-workgroup information access, cooperation and co-functioning. This is provided next and it too does not require justification.
3. Every information worker who functions regularly as a member of a specific workgroup, where the individual requirements of that worker and/or of other members of that group so indicate, has the right to be provided with client-server and/or multi-user capabilities. These include OA (TS = 2000) and/or application/package (TS = 3000), over and above application/language capability (TS = 1000). This is provided next and does not require specific justification. However, the cost of providing such service shall have deducted from it the cost of having first provided basic desktop and network connectivity services. Only the *marginal* cost of the capability to accommodate workgroup needs must be justified against the benefits of packages and applications to be run on such systems.

With such a 'User Bill of Rights' in place, we would first consider delivering resources to meet a User Demand Profile (UDP) of TR/CR/SR of 1000/0/1000 and, thence, the requirements of the higher profiles as invoked by user requirements and consequent exercise of rights by users who submit requests for service.

Alternatively, another variation of this approach could be utilized. If the CEO (or an appropriately empowered CIO) should mandate that each information worker is automatically entitled to the full OA service level (embracing both (1) and (2) above), this would mandate provision to each user of a, partially shared, capability of TS = 2000, plus a minimum CS = 1500 rating plus the thus-inflected SS support rating. Clearly, however, this sort of mandating is very expensive.

Surely you can put your accountants and systems planners to better use than having them read through reams of justification documents for systems which everyone already agrees are really needed anyway? Empower local line managers by giving them 'automatic credit' for the following:

- basic desktop PC capability
- desktop windowing environment of their own (or the organization's) choice
- local area network (LAN)
- MBS (or preferably MRS) to deliver OA services
- wide area network (WAN) connection

An RDBMS (and any corporately mandated universal applications) might be added to the above list. Then, let the local manager justify only the incremental requirements he or she may have, thus 'topping up' system capacity. If an OA-ready system is provided (with at least TS = 2000 per user) then the worst-case scenario is that of adding about TS = 1000 per user. If the corporation has already provided an RDBMS, then top-up may merely involve adding more main memory and disk, as well as any required specialized peripherals. This is the recommended approach.

Economic analysis framework

This section attempts to establish a basic framework or approach for the conduct of economic analysis for individual investment projects within an open systems environment. It is assumed that the reader's analytical approach will be a mixture of business case/cost justification, cost–opportunity–benefit analysis (COBA) and make-or-buy-analysis. The previous SFM, and an appreciation of the economic issues set out above, have been integrated into a single approach, which is referred to as the economic model of information technology (EMIT). It is also assumed to be important that the scale, and level of effort required to actually use, any analytic approach or tool be in concert with the project being considered. It is folly to spend $50 000 conducting COBA for a $250 000 project. It is not possible or practical to create a complete, ready-to-use, analysis tool here. However, this chapter does provide the basis for the establishment of an analytical approach which can meet the user's requirements with appropriate customization.

We first consider the basic motivators for an information management project (of any kind) and then address the project phases which are most relevant to EMIT, specifically:

- project initiation
- project classification
- requirements identification, including work item analysis
- determining investment objectives
- feasibility, including environment selection
- business case/cost justification/payback
- post-implementation review

Basic motivators

Why would any line manager ever want to become involved in an information management project in the first place? This is the fundamental question which must be borne in mind not only by line managers (and their end-users) but also by the professional informatics staffs who assist the manager in considering, planning and implementing such a project.

In principle, the line manager wants to maximize his or her unit or workgroup's efficiency in managing information. Most progressive organizations today mandate that information is to be managed as a resource, and on a life cycle basis. Advanced technology clearly has the potential to increase leverage in accomplishing this. However, it must also be considered that counter-balanced against this is the tendency for many end-users and line managers to view the professional IT field as a thicket of incomprehensible acronyms, rules, devices and causal interrelationships. Not to mention constant, sometimes almost frantically paced, change. Often, a manual or an existing (past generation technology) solution will be retained far past the economic and operational cross-over point (from which point forward the new solution is provably better) due to factors of inertia. A project will launch – and keep progressing – only when the initial perception, and the continued perception throughout the project life cycle, is that not only will marginal benefits exceed the marginal costs, but also that the project can be realized without unduly stressing the workgroup. Clearly this is a perception on the part of the intended end-users and their line management. The fact that the IT department thinks the project is wonderful means little if local end-user (and hence local line management) enthusiasm wanes. Local line management must not only continue to believe in the cost/benefit usefulness of the project, but also its continued 'doability' within the administrative and operational structure of the organization. The onset, or advent, of budget cuts, reduced or increased downsizing targets, the end of the fiscal year and other factors can influence this. There is thus a 'political economy of project doability' which must be borne in mind by the IT management and the combined IT-user project team.

Where specific business applications are concerned, the workgroup may want to obtain an application to fulfil one or more of the following functions:

- handle common business information and processes in which groups of end-users will share on a sequential and/or simultaneous basis
- permit new (and/or more) business information to be handled, thereby increasing productivity if all other things remain equal

The application may replace a manual system or a combination of manual systems and existing automated systems. It may be desirable to bring down to the workgroup level a process now performed on Area or Enterprise equipment where it is essentially a local process or is one which could benefit from local enhancement. In the alternative, the application may have started out on one or more PC platforms and may simply have outgrown the single-user (possibly even the client-server) model.

There is also the case of an application prepared by a corporate group at headquarters (such as Human Resources or Finance) which is then provided to the local workgroup with the expectation that they will run it. Where this is provided as a PC/MSW application, the group originating the application in most cases expects target user individuals to have a PC available. In future, such a central corporate department may similarly expect a workgroup to have a client-server (or even a full multi-user) system available. Where the application thus provided to the unit is nonetheless the first application to require such a platform, it will require provision of such a platform if it is not already present.

Project initiation phase

Where the workgroup (and its management) have concluded that an informatics project should be undertaken the first step is to produce a sufficiently clear statement of the objective, goals, bounds and scope of the anticipated project. At this stage the first intake of economic information for EMIT occurs. How many people are in the workgroup? What size of application and/or system is contemplated? What order of magnitude of benefits and costs are expected? The expected 'lie of the land' in terms of costs and benefits can be considered against the basic model in Fig. 4.2.

At this stage any factors which fall outside the standard or 'stock' variation of the model should be identified. It may be desirable to progress through the above-cited assumptions as a checklist in order to establish any anomalous or unique aspects of the workgroup, their requirements or mission or of the proposed project.

The degree of interaction that the workgroup has (and which the application/system is expected to have) with organizations, systems and individuals outside the organization should be documented at this stage. Some of these relationships may impose absolute limits on technical, economic or operational aspects of the application or system or even on the timing, staging and deployment of the project itself.

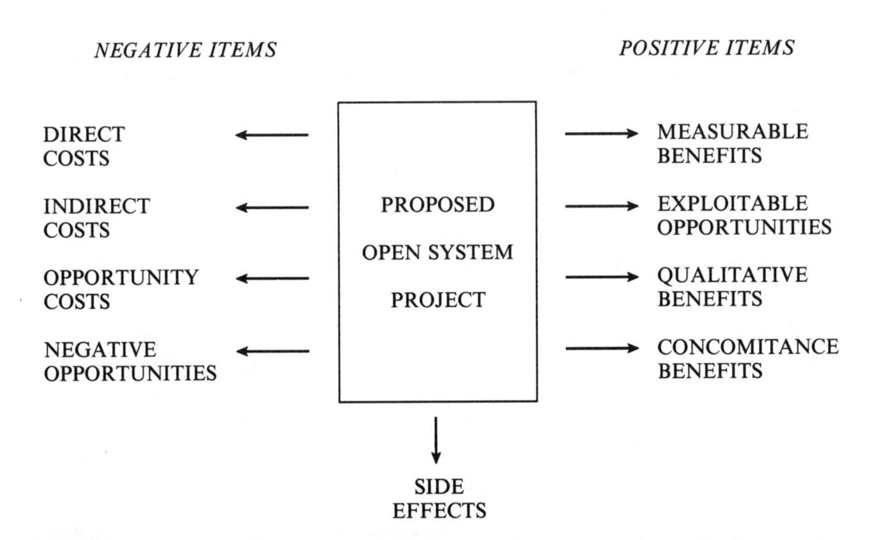

Figure 4.2 Project inputs and outputs.

Project classification

From the IT perspective the 'fish or fowl' question of whether this is expected to be a PLATFORM or a PROGRAM project is very important. The PLATFORM approach will place primary stress on somehow obtaining the best possible application for the workgroup and then finding a platform on which it can reside. It assumes dedication of the system as the platform for a sole application. The PROGRAM approach is much more broad, seeking instead to ensure that a general-purpose system with reasonable immediate surplus capacity (and expandability) is set in place to meet a complex of needs, of which the spurring application (where such exists) is not necessarily the most important or profound. In technical, operational and economic terms the PROGRAM mode is therefore a superset of the PLATFORM mode. Determining the PLATFORM or PROGRAM nature of the project should provide important information – even at this stage – as to which types of costs and benefits will be considered. At this point, it may also be desirable to make assumptions about what will be deemed to be installed first, where marginal costing of (for example) OA over application/RDBMS versus application/RDBMS over OA is concerned.

At this stage there are two generic options possible for the project team in seeking to understand, characterize and assess the work requirement of the workgroup, which the project is to address. These are:

- Conduct a conventional user requirements analysis as set out in the organization's standard system life cycle management document, where such exists – this will usually address data entities and flows as well as user functions and processes.
- Conduct an analysis of the workgroup work item (WI) or task processing requirement in the context of the relative place of the group of WIs comprising the process to be captured in the application/system as related to the universe of all WIs normally processed by that unit

(Appendix A contains information useful in specifying a model of the flow of work items within the workgroup and between the workgroup and other entities. The approach is quite generic and is believed practical for use within your organization. Note that neither EMIT nor the system functional model (SFM) are dependent upon the use of the WI approach model, but it will enhance both of them.)

From EMIT's point of view, if the specific *portion* of WIs undertaken in respect of the specific business process being automated can be compared in percentage terms to the total group of WIs now being undertaken, we can establish with reasonable accuracy the current total costs of the existing system or solution. This is because the total of all work items undertaken (annually) by the workgroup is supported completely, and only, by the directly allocated resources (and share of corporate overhead resources) provided to the workgroup by the organization. Thus $X supports a WI volume of A and $Y (where less than $X) supports WI volume B, that subset of the WI volume related to the particular business process(es) now being considered in the project. For this reason alone, the WI method is much superior to conventional data flow analysis or business process analysis because their various respective totalities can not be readily costed against total direct/indirect operating cost of operating the workgroup or unit.

With respect to the SFM, the WI method would permit Level 8 (the user level) for a workgroup to be characterized as a combination of manual, semi-automated and automated WIs which in virtually all cases will exceed the total capability of *any* installed combination of single-user, client-server and multi-user systems which supports that workgroup. Thus, Level 8 would (in TR or at least TR-equivalence terms) significantly 'overhang' the provided system capability contemplated by the project (Fig. 4.3). However, the amount of overhang would, especially in the case of a full multi-user system, undoubtedly be reduced significantly by the project. Of course, the project also contributes to increased costs for the workgroup or unit, and this is treated below. Figure 4.4 provides a more generalized appreciation of the relationship of the WI to its environment.

There may be configuration-limiting or other limiting factors that must be addressed at this point. Perhaps the unit has a mission-critical application which must be cut-over without disruption or which has other special handling requirements from the technical, operational and/or economic perspective. Perhaps some outside agency must be paid something with respect to the change, or will help pay for the change. At this point, any issues of revenue-generation or cost-recovery must be addressed.

Investment objectives

Whereas the project planning, including its analysis component, have come this far, it is now an appropriate time to determine basic investment objectives. If the investment only touches an application, are there any objectives beyond just obtaining the software? A system planning methodology which is application source neutral is required if the

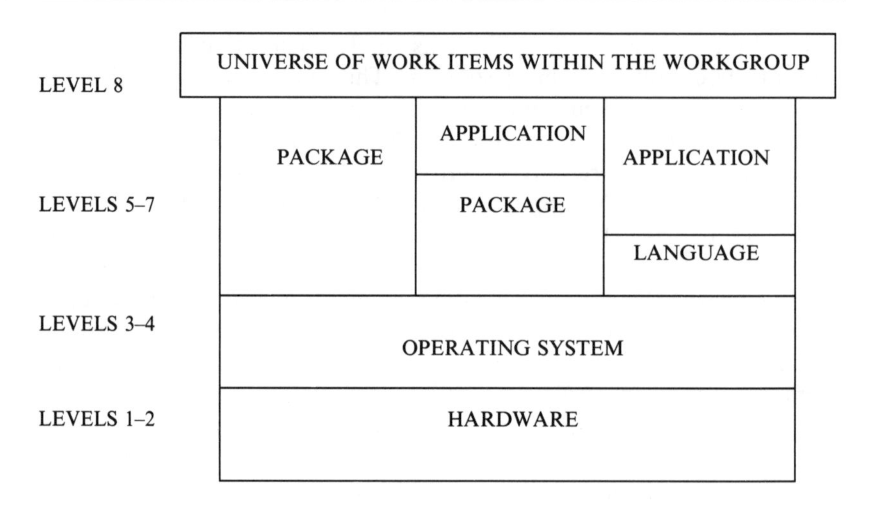

Figure 4.3 Work item (WI) level 8 or 'T model'.

(Source: See system functional model (SFM) in Chapter 3. See Appendix A for more information on WI methodology.)

Note that the portion of the work item (WI) universe not supported by the workgroup system is supported by manual processes and/or by other systems not modelled here. This diagram addresses only the technology supply (TS) stack, ignoring communications and support.

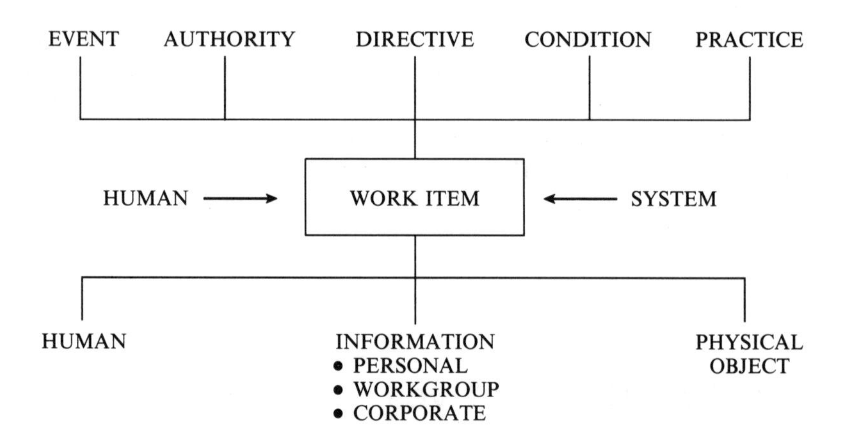

Figure 4.4 Work item (WI) environment.

Work items are spurred or created by various stimuli, are performed by systems and/or humans and they act upon humans, information and/or physical objects.

software registry concept is to be successfully implemented. Specifically, if the system life cycle methodology implicitly assumes that the user group will be drawn towards creating a new, custom application in most cases, then it will be biased in the functional element of its make/buy analysis. If the final choice is one with no application capital expenditure (which uses a pure REGISTRY application with no changes) this may heighten the relative importance of corollary corporate or workgroup investment objectives. If the application gives rise to a platform which is to be a *de novo* MBS/LAN or MRS/LAN configuration, it will bring with it a UNIX system administrator who will undoubtedly increase the quality – and immediacy of availability – of end-user *in situ* support. While obtaining better PC support may be a secondary objective of the investment, it may offer a significant payoff.

It is desirable at this stage to classify the objectives of the investment into the following three categories:

- PRIMARY: those objectives that must be met in order for the investment to be provably worthwhile and not trivial or counter-productive – we must *diligently seek* all costs and benefits which fall into, or relate directly to, this category.
- SECONDARY: those objectives that are clearly and obviously related to the investment, and the costs/benefits of which we will *seek* to understand and quantify/qualify where possible, but having regard to the scale of the project and thus the scale of the analytical effort required to support it.
- TERTIARY: those objectives that are incidental to the direction of travel of the project (or workgroup), and the costs and benefits of which we will in general measure only as they 'fall out' of the analysis of other technical, operational or economic issues – the only caveat here is that we may overlay a corporate template which lists such costs and benefits in a generic way.

There is a 'special case' of the above taxonomy of objectives which relates to the concept of basic (corporately mandated) OA entitlement as discussed above. Where some or all of a line manager's direct-report workers had not yet been provided such service, he or she could – in seeking to justify a platform for a given application – also 'call down' an automatic corporate justification to provide a basic level of OA service to all such workers (likely TS = 2000/CS = 1500 per the SFM). In such a case, only the *marginal* or *incremental* costs of technology, communications and support capability (as inflected by the first two) over and above the 'basic OA platform' would need to be shown to be offset by marginal benefits from that application.

☛ The OA issue is by no means a trivial one, since adoption of a corporate OA mandating structure would provide a very major boost to workgroup processor acquisition, and hence to the pace of migration to open systems. This would, in turn, open the possibility of more new (and re-developed) applications being offered an optimum choice of tier, assuming at last some UNIX capability at the midframe and mainframe tiers.

Feasibility phase – environment selection

At this stage it is also necessary to define the alternatives, some of which may have causal relationships to others at the same stage downstream. For example, a third-party application may be dependent upon an RDBMS not normally supported by IT. Worse still, it may require a proprietary multi-user system not now in use within the organization.

Once the alternatives and the initially known relationships have been set out, it is possible to proceed with cost–opportunity–benefit analysis (COBA). For each alternative, the core stream of costs and benefits (as impacted by other factors discussed paragraphs below) must be considered. In the purely PLATFORM mode of acquisition (wherein a general-purpose machine is purchased), we are looking at a volume/texture of total work items (WI) processed which represents the total work now done without the application and system being in place and also the total amount of work to be done when they are implemented. For more information see Appendix A. Where conventional requirements analysis is used (i.e. not the workflow analysis approach) the principal remains basically the same; we have a work objective to accomplish. The key measurable benefits of the application will be those which increase production and/or decrease required work time under the same timeliness and quality conditions.

If WI analysis is used, we may also want to assign some value to the time of various workers that is freed up for potentially more creative activities and which is not otherwise reallocated either to other existing production work nor consumed in support of downsizing (or rightsizing). The 'value of output' is not treated expressly here, but it can certainly be derived, particularly where the application/system increases productivity. This can be accomplished by fully costing the present versus projected (post-implementation) total of work items (WIs) processed. The problem, of course, is how to determine whether the additional work items possible in the post-implementation environment should be valued at the per-WI cost as it existed originally, had production not actually increased, or as its share of the new expanded production. In a market-driven situation, where the customer directly consumes the outputs of WIs (be they hamburgers, automobiles, answers to calls on a software support help-line etc.) it may be easier to assign a value to each WI and its attendant output since the market itself may determine these values.

The next most important benefit to be considered is the interoperability benefit that the implemented application and/or system will provide. The application itself may provide important new interoperability with other workgroups inside the organization and/or with outside customers which may bring savings not contemplated by the original framers of the project. Where the application begets a new platform, there will be platform-related interoperability benefits as well.

An important benefit from any UNIX-based system acquisition is the potential for benefits from data and particularly application portability. Leaving aside OA (with its usually inherent LAN and WAN access), and assuming the case in which obtaining the application also leads to acquisition of a platform, this leads to the potential for using one or more of the three axes of portability discussed earlier within the workgroup or among workgroups. Additional costs, opportunities and benefits are listed in Chapter 2 (and detailed in my previous book). So is the concept of the Registry and the concomitance of adoption of new technology. Chapter 3 of this book provides a stable framework for costing specific items of technology on a consistent and comparable basis.

One of the most vexing issues is how, within a large organization, to levy on individual workgroups the strategically related costs of establishing an open systems technology in-migration and support capability. These include planning, system integration (or at least oversight of system integration), user and system administrator support, network management and technology monitoring. One approach is to determine the core and variable (but non-discretionary, as in demand-driven) cost of providing IT functional support and other indivisible services and then fund these as corporate overhead. Many IT organizations function in this way, although the internal chargeback system is also widely used. Where outsourcing is a possibility, getting a firm handle on the market value of this (or any other) set of services provided by IT is necessary. For example, if a comprehensive open systems support capability is established by IT, it may be desired to charge each new workgroup migrating to UNIX-based MRS and/or MBS equipment $20K in the first year for this support and $10K for each subsequent year. In some accounting regimes, at least the first-year cost (the cost of helping the workgroup get started in open systems in the first place) can be capitalized, while the second and subsequent year support charges are treated as 'O&M'. This approach is recommended where practical, although it will distort some forms of make-or-buy analysis and plays havoc with the numbers when the equipment itself is to be leased from outside the organization. However, the analyst should note that where such a levy includes or helps cover any form of contract between IT and the technology provider (which reduces or eliminates the local site's O&M cost for such things as annual maintenance, software licences or upgrades), this cost must not be counted both as part of the broken-down local cost *and* as a part of the levy. Where it is included within a mandatory levy upon the workgroup (or a voluntary one which the unit accepts) this is the preferred means of treating the cost.

Where an application is being 'placed' within an existing technology infrastructure, there may be an opportunity cost once the application is placed, for example, on an MRS or MBS. It may then become impossible to install more OA or other software packages for other users or even more data for existing applications. When one or more of these limitations arises, two alternative remedies should be considered and costed:

- Accept the cost of degradation (including any resulting contention or other problems), but count it as part of the cost of the new application being implemented now – the degradation is, after all, caused by the arrival of the application.
- As part of the project to put in place the new application, also include a sufficient increase in TS and/or other capacity to return the system (after the new application is installed) to the same amount of reserve capacity as existed before the installation.

It can readily be seen that the SFM discussed in Chapter 3 provides a reasonably intuitive and quantitative means of accomplishing this. Where it is possible to use the SFM (or any other tool) to estimate them, the costs to resolve any contention (which is preemptively resolvable at implementation time by buying more technology then) should be included as a cost of the project.

While your own organization's costing assumptions and approaches may differ, and these should in any event be varied where there is a project-specific justification for doing so, the following standards or default set of assumptions can be utilized as a fallback:

- inflation at two or five percent
- taxes and governmental licence fees at actual pre-tax cost
- salary rise at your organization's average rate of increase
- five percent nominal discount rate
- five year project life cycle horizon, but with concentration on a PRIME *period* of the first two or three years, during which full, or at least a substantial portion, of payback should occur

- a blended 'takeout' of price/performance improvements over time, as reflected in components or upgrades purchased later, such that we will take part of the technology gains in improved performance for the price and part in reduced price for similar performance

In most cases, it will be quite difficult to estimate the cost of a security breach arising from or caused by the new application and/or system. The obvious exception is the application which brings with it (or at least admits to the organization) a virus which crashes the local workgroup (MRS or MBS) and PC environments and has clearly identifiable disruption, down time and recovery costs. In some cases, the local line manager may be able to furnish the IT analyst with a reasonable estimate for the costs of other types of security breach.

The potential for inter-tier processing power trade-offs requires development of a more temporally sensitive versions of the three-axis portability model of Chapter 2. It is almost certainly a problem requiring a rigorously quantitative approach. Certainly, some systems (when staged or deployed in certain ways) offer a higher potential for trade-offs than do others; trade-offs are easier when the systems are planned, procured and installed at the same time. Where measurable savings in total TS delivered to a workgroup can be identified and measured, the resultant (directly associated) capital savings would be a result of such trade-offs. In most cases, such trade-offs will be among PC, WS, MBS and MRS equipment, although trade-offs among larger classes of equipment are also possible.

☛ Sometimes, buying bigger PCs or workstations lets you buy a smaller MBS or MRS, particularly if they both run UNIX. Conversely, if you buy a bigger MRS you may be able to buy less costly desktop equipment.

Regarding the comparison of the benefits provided to individuals (versus workgroup) by a new application and/or system, it is recommended that the WI model set out in Appendix A be used to help identify those opportunities for work and workspace sharing (including co-editing and co-processing) which have time-saving and/or productivity improvement potentials. In most cases, the selection made between the client-server and multi-user service models governs the degree of workspace sharing which is possible. If opportunities to use a shared workspace are valuable to the workgroup, then presumably system configurations which present such opportunities to a greater degree are more valuable. Fundamentally, however, the line manager and the end-users themselves must be able to impart to the IT analyst some information which permits quantification of the benefits of such sharing. The WI model will tend to facilitate these discussions by focusing everyone on the classes of work items which connote – or at least might benefit from – true workspace sharing. The total annual value of these work item classes gives a baseline for determining marginal or incremental benefit, or at worst producing a ball-park estimate. In home and remote workplace situations, this is a crucial factor.

☛ If open systems let people more easily share their work, will the tasks they share be accomplished more quickly or be completed to a higher standard of quality after the new program and/or computer is installed? How important are these tasks?

The cost of disruption related to a configuration-induced failure should be determined by seeking to find out the probability of such a failure and rating it against the estimated costs of the failure. This is a particularly important issue where local 'cheap and cheerful' advocates loudly advocate purchase of low-cost $10–20K MBS equipment over an MRS which may cost three or four times as much, but will not be in danger of quickly becoming over-capacitated. Server cloning (when the MBS equipment runs out of steam), in an RDBMS environment, is often an open invitation to configuration-induced failure.

☛ If you buy a system which is probably too small, you should estimate the probability-weighted cost of it failing and forcing you into an unplanned, unscheduled and disruptive upgrade.

There is also the issue of costing a configuration-induced involuntary upgrade itself (whether it be MBS cloning or upgrade to an MRS from an MBS-only). It can be treated similarly by using the above-derived probability weighting against the likely cost (in current real dollars) of the upgrade.

☛ You should also cost out the upgrade itself if there is a reasonable chance you will have to make such an upgrade solely because you did not buy a big enough system in the first place.

At the conclusion of the feasibility stage, we should have determined:

- the best means of obtaining the desired application functionality, and/or
- the best means of providing a platform for general usage or for a specific application.

Business case, cost justification and payback

Once the COBA is complete, and all other non-cost factors have also been considered, the line manager and the IT project team can together select the best overall project approach and can proceed to develop a formal business case to support it. This will include full cost-justification of the proposed original investment and ongoing operating expenses in terms of a monetary payback. Most often, constant values are used for this analysis; therefore, current-year values and then-year values are treated differently.

Post-implementation review

Whether the project is a glowing success or a dismal failure, it is important always to undertake a post-implementation review. The economic element of such review should address the following question about costs:

- How did actual costs compare with the forecast?
- Where costs diverged (either up or down), was the change due to
  - estimating error?
  - price changes?
  - purchase of something other than what was planned?
  - changes in taxes (almost always up) or handling costs?
  - unconsidered or unexpected items?
  - breakage, loss, sabotage or theft?

Regarding opportunities, it is necessary to try to identify which of the forecast opportunities have materialized or been exploited and, where possible, to quantify the beneficial and non-beneficial outcomes. In some cases, it will be necessary to consider the positive and negative aspects of a given opportunity solely within the context of that opportunity, determine the 'net' outcome and then quantify its implications in terms of a cost or a benefit, as the case may be.

The key questions about pure benefits are:

- Did the benefits materialize when and to the degree expected?
- Where the benefits diverged (up or down) from those forecast, was the change due to:
  - more users than expected but still within system capacity?
  - more types of uses made of the system?
  - more work (WI volume) handled?
  - better system performance or reliability?
  - better user learning retention (of training)?
  - application and/or system put to unplanned use or enabled an unplanned application or system?

Conclusion

This chapter provides a much more vague set of guideposts than Chapter 3 (which covers the SFM) or Chapter 5 (on the application placement methodology). More than a topographical map of how to navigate that terrain called economic analysis, it is a crude plot of the minefield which also exists there. A comparative review which I undertook of the progress of open systems in North America and Europe made clear that the latter's much better progress was due in part to a greater emphasis on strategic thinking on the part of senior management and government leaders. Short-term 'this quarter' thinking seldom favours a strategic commitment to anything as difficult and potentially hazardous as open systems.

Many is the time that the best formulated technical plans have been thwarted by the Finance Department. The SFM and the COB list provide good tools to ensure that this does not happen, but only if they are judiciously utilized. The best tank, equipped with the finest gun, can always be damaged or destroyed by a land mine or just driven into a deep ditch! Remember always to refer back to the high-level linkage which you earlier established between open systems and business objectives. If all else fails, let me provide you with a 'secret weapon'. I was both amused and pleased to note recently that many European governmental (and some private) organizations have now begun to link open systems directly to the quality process, and even to the ISO 9000 standards. This presents a very effective means of smoking out and extinguishing the last vestiges of internal opposition to open systems: let the manager (even the financial manager) who is not in favour of quality please stand up!

5 Application placement methodology: putting applications and technology in their place

Introduction

This chapter addresses the issue of how to achieve optimum application placement among tiers within a four-tier system architecture, which is usually the highest number of fully distinct tiers found in a given organization. It is readily adaptable to a three-tier environment, but would be distinct overkill in a two-tier situation. First, a series of assumptions are set out and then application placement drivers or factors are cited. The assumptions are divided into the following sections:

- the environment of a workgroup
- work and workflow
- system architecture
- development/implementation process
- the application itself

Finally, a rule-based method is proposed, called the application placement methodology (APM).

Basic assumptions

1. The organization has developed a mission statement which is relevant to all workgroups, particularly those who will be among the vanguard in acquiring new technology. All applications must demonstrably contribute to the individual or workgroup ability to further this mission. The workgroup has a well-defined business mission and clearly defined intra- and extra-workgroup work relationships.

2. The following general constraints may attend consideration of where to place the application:

   - Placement on more localized tiers (workgroup or workplace) may not be feasible solely because the workgroup is at a point at which it cannot tolerate the disruption of the implementation *both* of the application and one or more new processing platforms required to host it.
   - Budgetary constraints may preclude the use of processing tiers which would otherwise be available.
   - The 'political economy of project doability'.
   - Management comprehension of the requirement and/or the differences among potential solutions (e.g. some managements may simply not believe in the benefits of OA).
   - End-user sophistication does not permit a given architectural solution (e.g. feasibility stage of project recommends UNIX workstations on all desktops, but users now have 50% non-intelligent terminals and 50% no desktop device at all).
   - Relationship with external client dictates use or avoidance of specific platform or architectural approach.
   - Security requirements dictate or preclude certain approach.
   - Other manual or automated systems support (or are supported by) the work item grouping represented by the planned application and must be modified or replaced at the same time as the application goes into production.
   - User desires are seen to conflict with principles of good system deployment or with system architecture (e.g. it may not be operationally desirable to place a 'corporate' processor on an individual's desk even where this is technically feasible).
   - IT workload may impact the pace of the project through the phases set out earlier – if the pace is perceived as too slow the unit may seek other temporary or even permanent solutions, probably also seeking to cancel the project.
   - Inability of suppliers to provide software, hardware or services when promised (and on an as-promised, complete basis) may reduce the pace of or halt project implementation – some processing tiers may be more subject to such constraints than others.

3. There are three alternative deployment strategies, each of which has positive and negative aspects. It is essential that IT and the unit agree upon (and the unit 'buy into') the intended strategy at the end of the feasibility stage, when it is known where the application will come from, what (if anything) will be done to it on its way to the unit, where

the hosting platform will come from and what service model and I/O strategy are to be employed. The three alternatives are:

- first develop a *prototype*, *then* move to *production*
- first develop a *prototype*, *then* run a full scale *pilot*, *then* make any further modifications necessary to move to *production*
- develop the application, accept it and move *straight to production*

Note that these options specifically refer to the DEVELOP application acquisition strategy of the Registry, but can be inferred to apply to the MODIFY and THIRD PARTY options as well. Note also that production may commence at one or all sites on a given day or may be implemented on a cascade/progressive basis, as experience is gained.

4. It may be possible to use the work flow methodology (WFM) to assist in identifying opportunities to improve quality assurance within the conduct of the part of the unit's business to be impacted by the new application. See below and also Appendix A.

Work/workflow assumptions

1. One alternative for assessing organizational and workgroup workflow is the workflow methodology (WFM) which is presented in Appendix A. This approach can be used for estimation of opportunities for improved efficiency and/or improved creativity when a potential new application is proposed.
2. Even if conventional data flow and process (functional) analysis is utilized as one of the backbone elements of your system development life cycle, it will still be possible to estimate with high confidence the percentage that the work items embraced in a new application (here called 'WI(A)') represent of the total work item universe (here called 'WI(T)') of the unit on an 'at-par' basis, without yet considering the benefits of automation in improving worker task efficiency. The WI(A)/WI(T) ratio also represents the degree of importance of the application to the unit, and therefore probably also its priority, at least to the unit (see Figure 4.3).
3. Most (or all) of the work items (WIs) to be performed by the application, either on its own or in concert with other applications and/or end-users, can be classified under each of the following:

   - intra- or extra-unit in nature
   - relating primarily to the Workplace, Workgroup, Area or Enterprise tier of information management
   - of single-user or workgroup concern

- requiring either a single-user dedicated (or else a shared) workspace
- requiring either a single-user to perform or else a workgroup to perform each of the given generic functions upon the WI
- with or without significant potential for use of the Smart Work Item (SWI) paradigm (see item 5 below)

4. Within a given WI grouping, any given WI may, as a result of the implementation of the new application, emerge in any of the following states:

   - still performed manually
   - performed partially on a manual basis and partially on an automated basis
   - performed wholly on an automated basis

5. When a WI is fully automated it may be characterized as:

   - non-intelligent (submitted in a batch-like manner and simply performed)
   - semi-intelligent (the work item itself, not the application but the total work item, can be seen to execute 'IF-THEN' and similar basic logic) – it carries a 'business method' with it
   - fully intelligent (utilizing artificial intelligence capabilities of the system – in such cases it can correctly be seen as an SWI)

   Note that further delineation of the boundaries between these three classes of WI is necessary before this assumption can be fully operative. This touches both work analysis and system issues, including a formal statement of the types of intelligence which an SWI can reasonably exhibit.
6. The new application will respect (and hopefully improve) the existing WI flow, prioritization and interrelationships, at least within the unit or workgroup. In certain cases only a portion of the workflow model (or a subset of the types of human/machine functions) will be required to consider the WI grouping addressed by the application.
7. Where practical and beneficial, comprehensive workplace simulation (CWPS), as in Appendix A, or a similar approach, may be utilized to assess the pre- and projected post-implementation flows of WIs. Use of such a tool permits the analyst to vary any of the following:

   - WI grouping addressed
   - WI flow and prioritization
   - data addressed by one or more WIs
   - technology to be provided
   - number/role of workers

8. CWPS (or a similar approach) will also permit assessment of the total resource consumption implications of the various combinations of factors under item 6 above. The total savings potential (or the nature of the savings), compared with the current or baseline situation, may differ under different workflow scenarios – even with exactly the same technology applied. It must be recognized, however, that experimentation with the workflow itself is experimentation with how the work is done, by whom (also by what system), where, when and under what conditions. Variation of these factors is very much an issue of how the unit or workgroup does its business. The local line management, and end-user community, must therefore be full participants in the scenarios to be developed. There is no point in discovering that modification X to the workflow permits the application to reside on a cost-efficient and otherwise appropriate tier of processing if such modification is not operationally practical.

The clear benefit is that this analysis may identify important opportunities to reorient the workflow to be addressed by the application and/or the information managed by the application, in a way that the workgroup may find acceptable, *before* finalizing the requirement statement (and hence casting the mould for the application). Besides identifying inefficiencies, the process may also identify anomalies, inconsistencies or other problems in how the work flows currently and/or in how information is produced and managed.

9. A crucial reason to use CWPS or a similar approach is that it will identify the 'who, how much, when and where' details of new envelopes of free time (time saved) brought about by the implementation of the projected new application. These will usually be tier-independent (they will occur no matter what size of system the application runs on), but there is no assurance that this is always the case. Analysis under item 6 above should indicate any tier-dependence that exists.

10. In most cases the priority of the majority of the WIs (or at least the crucial WIs) being addressed by an application will, in turn, determine the priority of the application in terms of speed of implementation, redundancy, security and similar issues.

11. It is necessary to identify *all* of the unit's WI relationships to external entities, specifically:

- WI inflow to and outflow from the unit
- WI co-functioning by persons or processes inside and outside the unit

- WI interdependencies, interrelationships and interoperation with WIs performed outside the unit
- prioritization parallaxes (WI X is dependent upon external WI Y, but Y has a lower priority and security rating than X and therefore represents a potential Achilles' heel to the workflow)

System architecture assumptions

1. The organization will adopt a unified architectural approach, and system implementation (in its technology selection and deployment) will therefore be architecture-driven and not merely technology- or vendor-driven.
2. IT functional authority will be fully understood, respected and heeded by line managers. They will solicit and will follow functional guidance and advice tendered them by or on behalf of the CIO. Requirements for expediency in implementation, preference for a specific vendor and similar factors will *not* be used as arguments that functional advice should not be complied with. (This assumption may be too heroic for some organizations, particularly those where some user groups are both technically sophisticated and politically powerful. Such groups may covertly or even openly challenge the IT Group's authority to set and enforce an open systems strategy.)
3. The system architecture adopted will be capable of being represented (in words, diagrams and dynamic (animated) graphics) on a fully comprehensible basis to central and local line management and to end-users who are not informatics professionals. There will be general understanding of, and agreement upon, the assumptions, procedures, terms and definitions inherent in the system architecture and SDLC.
4. The organization is committed to open systems, will deploy UNIX at all tiers when possible and intends to use OSI, or at least TCP/IP, networking.
5. It will be possible to map the existing technology, of the user workgroup, against the system architecture in a graphic and user-understandable fashion.
6. The System Functional Model (SFM) can be used to demonstrate the demand for, and supply of, information technology resources using elements of the triple stack model (technology, communications and support) discussed in Chapter 3. Where the model identifies real or potential capacity bottlenecks in any of the three stacks, these will be addressed at the planning stage.

7. It is recognized that the SFM must be calibrated (and then re-calibrated at regular intervals) to take account of the evolving technology base.
8. IT will insist that any installed system have an absolute minimum of 30–40% surplus capacity in key resource areas (CPU, memory and disk) at the outset of system implementation.
9. PROGRAM and PLATFORM implementations will be treated as identically as possible from an architectural perspective.
10. A wide area network (WAN) generally incurs higher traffic costs (in terms of variable traffic-carrying costs) than does a local area network (LAN). Therefore, *ceteris paribus*, it is preferable, for example, to have application–OA interaction on the same platform, or at least at the same facility – versus over a long-distance leased channel – except where there is a clear and justifiable operational requirement to the contrary.
11. As UNIX is progressively implemented within each tier, the three axes of portability (vertical, horizontal and forward) will become increasingly practical and accessible to application developers, managers and end-users. A software registry will be established and will assist in this regard.
12. The following default 'maxims' will be utilized in application placement.

    (1) Place the application at the lowest practical tier (i.e. as close to the end-users as possible), all other things being equal.
    (2) The application will conform to the system architecture to the greatest degree possible:

- assume all architectural defaults (processing tiers, GUI, UNIX, POSIX, OSI etc.)
- the application will normally serve a workgroup
- information will be managed as a corporate resource
- the application will be sourced from and/or considered as a candidate for inclusion in the registry
- the organization may mandate a default RDBMS, OA package and GUI
- all applications which require at least basic security eventually will move to NCSC C2 Level of Trust
- as a default, it will be assumed that at the outset there will be NO requirement for:
  - remote access to the application
  - transaction processing OS or similar performance levels
  - parallel processing

- vertical integration of the application
- distributed processing

(3) There will be a reverse onus of proof on the application project team (local client staff, IT staff and any external contractors together) to establish to the satisfaction of a formal Project Review Committee (PRC) – or whatever such review mechanism your organization employs – that any of the normal defaults (as in item (2) above) should be overturned. For example, if the (third-party) application which the project team has concluded best meets the unit's requirements is not available under UNIX and without the required OSI and GUI components, the project team must clearly establish that:

- conversion of an existing architecturally compliant registry application is not practical or economic
- it is not practical or economic to use the third party application merely as a prototype for creation of a new application
- it is impractical or impossible to induce the application vendor to offer a conformant version of the application

(4) The unit or workgroup has absolute discretion in matters of fact regarding the nature of its business (including WI details and WI flow) and the business case to be made for the application and its platform, where relevant), but the various functional authorities (including IT) have absolute discretion in matters of fact in their respective areas of technical competence (technology, personnel, contracting etc.).

Development/system integration process assumptions

1. The project phases set out in Chapter 4 will be utilized.
2. Any one or more of the following may exist at the time the project commences (or recommences, after some discontinuity or lengthy interval):

   - the former application that the proposed application will replace
   - a functioning prototype for the proposed application
   - a pilot application actually in production
   - a similar application to the proposed application, located in the registry only and/or in production in another workgroup or elsewhere

3. The economic and context assumptions will be as set out in Chapter 3.
4. At the feasibility stage, the technology selection process will be architecture-driven except to the extent that the project team, as in the previous section's maxims 2 and 3, successfully argues otherwise. There will be an emphasis on exploiting open computing system scalability, interoperability and portability benefits.
5. Development, acceptance and production can all occur on different processing tiers. For example, development could occur on a UNIX workstation with deployment (acceptance/production) occurring on an MRS. Conversely, development could occur on a mainframe with acceptance on a midframe and production on an MRS or MBS.
6. Maximum use will be made of the registry. All of the REGISTRY, MODIFY, THIRD PARTY and DEVELOP alternatives will be considered. All other things being equal, the alternatives will be considered in order of least to greatest cost. This will usually (although it may not always) be the same order as set out above.
7. The SDLC will be (and will also be seen to be) source-neutral and therefore biased neither in favour of, nor against, any application procurement method. The SDLC will also be entirely neutral in terms of processing tier. This quality is also known as 'tier-neutrality'.
8. There will be cases in which the application selection process may impact the choice of RDBMS, operating system or hardware (type, scale or even vendor).
9. The application placement methodology (APM) will be used only *after* the feasibility phase first addresses the basic sourcing of the proposed application software. In some cases, however there may be a loop-back to the application sourcing phase. For example, there is the case in which the lowest cost (e.g. a THIRD-PARTY) application requires an MRS twice as large as would all other viable software candidates, thereby negating the cost of any saving on software. In such a case, application sourcing would probably be revisited.
10. The APM must be neutral to the selection (at the outset or during the earlier part of the feasibility phase) of the PROGRAM or the PLATFORM mode of technology supply.
11. The selection of the optimum combination of candidate software package (from among the four generic sources and from among individual candidates identified from each source) and system environment will be accomplished, in part, by using the SFM and the economic model of information technology (EMIT).

12. There is a trade-off in analytic accuracy in either:

- running the full SFM/EMIT for each application source/platform combination believed reasonably possible, or
- using a possibility tree or other approach to select only the most promising combinations for full SFM/EMIT analysis

The first approach is more thorough; the second may be more efficient. It should also be added that as experience is gained with the SFM/EMIT tools (not only in comparing head-end analyses but also in comparing these to their respective post-implementation reviews) a better ability to determine the degree of analysis necessary for a given project will develop on the part of experienced system integration and technology staff within the IT organization.

Assumptions about the application itself

1. The application must be *describable* (in system terms) as to the:

   - system resource requirements (processing, memory, storage, I/O and similar) – and hence describable in terms of TR under the SFM as imposed on lower tiers of that model
   - data to be managed
   - processes (in terms of WIs or in terms of conventional data flow diagrams and process diagrams)

2. The application must be *discrete* (in system terms), having clear boundaries or limits in terms of WIs, data, TS utilization and temporal (life cycle) issues.
3. The application must be *distinct* (in information management terms) from all other applications now in service within the organization, with the exception of an application sourced solely from the software registry.
4. There exist various *manifestations* of the application, some of which traverse the technology, operational and economic fields. These include:

   - original intended purpose or use (as-built)
   - actual purpose or role of the application (as-used)
   - system resource requirements (TR)
   - software acquisition, modification or development cost
   - time constraints

- interrelationships with other applications (horizontal and vertical)
- single-user, client-server or multi-user nature of service model
- standalone, LAN-dependent, WAN-dependent or LAN/WAN-dependent
- independent or package (e.g. RDBMS) dependent
- unilingual, bilingual or multilingual
- amenable to single or multi-processing implementation
- security rating (ultimately NCSC Orange Book Level of Trust)

Note that the as-built and as-used manifestations may differ. Perhaps an application designed to issue licences in one unit can be used by another for periodic licence reviews. This might connote expansion of the data to be stored.

5. Selection of an application service model connotes consideration of the various trade-offs discussed earlier, particularly when the choice is client-server versus multi-user.
6. When an application requires use of a non-standard package (such as a non-standard RDBMS), it must be held to bear the consequences (in the technical, operational and economic analysis) of that dependency. For example, if this forces a unit to acquire and dedicate a multi-user system to this application, and have a separate one for other applications under ORACLE plus OA (when this would not otherwise be the case), this consequence must be attributed to the particular application.
7. Similarly, when an application requires the use of a given operating system it must accept the consequences in the analysis. In general, indeed except where it is expressly proven otherwise in specific circumstances:

- UNIX is held to be more capable than all permutations of OS/2 and Windows NT
- Windows NT is believed to be more capable than OS/2
- OS/2 is held to be more capable than MS-DOS

Application placement drivers

The purpose of this section is to set out those factors which are at this time known to have an impact on the application placement decision, as related to a four processing tiers structure. In certain cases, they may also impact the choice between the client-server and multi-user service models. These factors are also known as 'application placement drivers'.

Specifically, an application placement driver is defined as:

a factor which alone, or in combination with one or more other such factors, impacts the placement of an application by encouraging movement towards, or away from, a given tier of processing or by explicitly including or excluding a specific tier.

Three types of application placement drivers have been identified:

- ABSOLUTE (ABS): forces or precludes placement of the application on a specific processing tier
- RELATIVE (REL): exhibits a tendency to force (or at least encourage) placement on a higher (*up-tier*) or lower (*down-tier*) processing tier (where such a tendency may transit the application one or more levels up or down), or which forces the application, user and/or data to converge towards or onto a tier (*convergent*)
- MINIMAL (MIN): has an impact which is perceptible (is not tier-neutral) and must be considered as a valid factor but is normally used only as a 'tie-breaker' or qualitative factor in placement decisions not otherwise resolvable by considering all ABS and REL factors

The ABS and REL factors are the focus of this chapter.

Application placement drivers can also be classified according to the element of the information management environment to which they most closely relate. These elements include:

1. The Unit or Workgroup
2. Work items and Work Flow Methodology
3. System architecture
4. Development and system integration process
5. The application
6. Economic issues
7. Other issues

The following terminology is used in classifying individual factors:

UP	for any *up-tier* factor of the REL type
DOWN	for any *down-tier* factor of the REL type
CON	for any *convergent* factor of the REL type
LAT	for any lateral factor of the REL type (where the factor induces a requirement for lateral processing or interoperability)

The tier values are as follows:

Corporate	4	(a.k.a. Enterprise)
Divisional	3	(a.k.a. Area or Regional or Divisional)
Workgroup	2	(a.k.a. Unit)
Workplace	1	(a.k.a. Personal or Desktop)

The following abbreviations are used throughout the data records in Appendix B, grouped by the various types of items:

SVCEMDL	Service model
ACQMDL	Acquisition model
SU	Single-user
CS	Client-server
MU	Multi-user
WI	Work Item
WP	Workplace
PC	Personal computer
WS	Workstation
MBS	Micro-based system
MRS	Midrange system
MDF	Midframe
MF	Mainframe
APPL	Application
REG	REGISTRY software sourcing option
MOD	MODIFY registry software as a sourcing option
TPY	THIRD-PARTY software as a sourcing option
DEV	DEVELOP software from scratch as a sourcing option
DAL	Default application locus
SCE	SOURCE
TGT	Target
TIER(DAL)	Tier of DAL
TIER(TGT)	Tier of target system
ABS	ABSOLUTE
REL	RELATIVE
MIN	Minimal or Minimum (also 'minimize!')
MAX	Maximal or Maximum (also 'maximize!')
XNN	Not NN
!NN	Must be NN
IM	Information management
IT	Information technology

When you can obtain an application from any of four generic sources, REGISTRY, MODIFY, THIRD-PARTY or DEVELOP, and can put it on a processing platform of virtually any size, anywhere in the organization, there is a tendency to lose one's bearings. In the 'good old days' it was rather more simple. PC applications came shrink-wrapped and minicomputer applications and mainframe applications came from third-party vendors or else were developed in-house or on contract.

☛ You need a methodology or set of rules to help decide where to put a new computer program, since you are no longer restricted and can put many new programs on any desired size of computer.

As a minimum, the following factors should be considered when seeking to decide on what class of machine to place a non-obvious application. Each one includes information on whether the nature of its impact is usually ABS, REL or MIN, plus any additional relevant information.

The unit or workgroup

1.1 There is an *external relationship* such that the current application is tied to a specific other application or process which cannot be disrupted by the implementation of the application, nor by the subsequent movement of the application to a lower tier. (REL) (UP)

1.2 There is a *budgetary constraint* which forces the workgroup to remain with the as-installed system environment. The workgroup can afford neither a new nor an upgraded system; nor can it afford to pay (presumably fully cost-recuperative) user charges levied for larger systems. The workgroup is therefore confined to its current equipment. (ABS) (!WS, !MBS or !MRS)

1.3 There is a *project 'doability'* constraint, apart from budgetary ones, which prevents use of anything but the existing system. (ABS) (!WS, !MBS or !MRS)

☛ Note that the degree of management comprehension, and the degree of user sophistication, are *not* placement factors as such. They are barriers that can be overcome, in most cases, without significant capital or O&M expenditure on the part of the workgroup or IT. Clearly, IT almost always has a mandate and a responsibility to provide orientation, training and education with respect to open systems.

1.4 Where the workgroup is not an organizationally and geographically discrete unit, the tier positioning or *tier residency of the workgroup* itself (and/or the key information it manipulates) will impact application placement. (ABS)

This 'tier positioning' of the workers brings into play three distinct but clearly related hierarchies:

- the organizational hierarchy of tiers (worker, workgroup, department or branch, corporate etc.)

- the four-tier IM hierarchy as it relates to the actual handling and management of information resources (workplace, work-group, divisional or area and corporate/enterprise tiers of IM focus)
- the four-tier hierarchy of platforms as it relates to the usual or customary (but not the only possible) manner of deploying technology, as follows:

Workplace	PC or WS
Workgroup	MBS or MRS
Divisional	Midframe
Corporate	Mainframe

Here we are primarily interested in the relationship between the first and the second hierarchy. Recall that the relationship between the second and the third hierarchies is not absolute. A workgroup with very intensive computing requirements might employ a small midframe as their workgroup system. Similarly, an individual managing a very large personal database could require a personal system with MRS-grade data storage and seek/access times. Someone whose work was devoted primarily to computer simulation might not need as much storage, but might require MRS-class raw processing power. Or, an enterprise level application might be given its very own dedicated MRS, but located inside the data centre.

Stated more formally, the average values of the tier positioning indicators of user workplace definitions and/or WI-impacted data items, where the tiers are accorded values of 1 to 4 as above, should provide an approximate estimate of the 'locus' of a given application. Of course, such an estimate may not be purely modal, perhaps being interposed between tiers (e.g. 2.75 is closer to a divisional than to a workgroup system); it therefore requires some interpretation. In most cases, the estimates derived from an analysis of workflow and data/information residency should not differ substantially.

1.5 In some cases there may be an investment *objective of bringing processing closer* to the workers or end-users. This might also involve invoking any corporately mandated OA funding scheme, as discussed in Chapter 4, to justify MBS or MRS equipment as well as the OA application itself. (REL) (DOWN)

1.6 The nature/type, standards adherence profile, approximate TS rating and age (and hence remaining service life) of any existing equipment (be it single-user, client-server or multi-user) will impact the application placement decision where such equipment offers the potential of serving as the target system. This may relate both to sunk costs and stuck costs as defined above. (REL)

1.7 The opinions or qualitative *views of unit management* and/or *end-users* (particularly any sophisticated end-users) may impact the favour or disfavour with which one or more of the make, modify, buy or get-free software sourcing options are viewed. This, in turn, may influence all remaining elements of the analysis in favour of, or against, a candidate software solution. Where this solution has system and/or tier implications, such views or opinions will (via the process of application selection) impact tier placement. (ABS)

1.8 The *location* of a unit's *worksite* may influence the choice of tier. All other things being equal, site locations which IT staff, and technology suppliers, cannot easily support will engender approval of higher tiers. (REL) (UP)

1.9 Where the *workgroup* is in fact *replicative* (there are many workgroups, each distinct and each replicating the requirement for the application) there will be a tendency to push tier selection down to at least the unit level. It may be necessary to consider the minimum, median, maximum, mean and mode of site size as measured by:

- total *employees* at site (ultimate potential application and/or OA user community)
- total *system users* at site (near-term potential application and/or OA user community)
- total intended *application users* at site (potential application user community)
- *current participants* in manual system and/or users of current application at site (immediate application user community) (REL) (DOWN)

Of course, each of the above measures of site population can only be compared with the same measure across sites.

1.10 Where all or most members of the *workgroup* either are *not* now and/or will not in future be *co-located* at one site or facility there will be a tendency to employ higher tiers of processing, particularly where workspace and/or function sharing is required. (REL) (UP)

Work items and workflow methodology

2.1 A ratio of work items (WIs) included in the proposed application to the total unit or workgroup WI universe (the RATIO OF WI(A)/WI(T), where A represents the projected application and

T is the total unit) can often be established. This ratio will in most cases govern the priority – at least from this workgroup's perspective – of the application. This will have at least three distinct manifestations:

- total system availability and reliability (REL) (UP)
- hardware availability, reliability and redundancy (REL) (UP)
- communications consistency and reliability (REL) (DOWN)

2.2 Each *individual* WI considered will usually have a clear *relationship to a given* IM *tier*. For example, a staffing action related to regional Personnel operations will probably relate to the Divisional tier rather than the workgroup or workplace tier. (ABS)

2.3 The workgroup may have a requirement for a true *shared versus single-user* workspace. (ABS) (XWS, XMBS, X all client-server)

2.4 The workgroup may have a requirement for true *shared* workgroup *functioning* within a given WI. (ABS) (XWS, XMBS, X all client-server)

2.5 In some cases the actual WI *flow sequencing* among workplaces (where sequencing options exist) may favour or disfavour a given platform tier. (ABS)

2.6 At the individual WI level, it is possible to address the impact which the intended *performance* of individual *generic functions*, at workplaces by workers, upon WIs may have on tier selection. Refer to Appendix A for a description of the workflow methodology (WFM). Note that direct machine participation in the generic functions of thinking, evaluating and travelling is excluded by the definitions contained in the WFM approach. The following are believed to be the tier implications of the remaining generic functions:

- *Analysing*: analytical functions are generally process-intensive and demand more TS capacity. (REL) (UP)
- *Comparing/contrasting/sorting/allocating*: these functions are less process-intensive and therefore more tier-neutral, but they *are* usually data-intensive – all else being equal, they should be located on the same (or at least the contiguous) tier to the key data sources upon which they rely. (REL) (CON)
- *Searching/referencing*: see previous item. (REL) (CON)
- *Reading/reviewing*: see previous item, but there is also the issue of shared review which may be impacted by the choice of service model (client-server or multi-user) – this is particularly true where the application must interoperate with the OA package. (REL) (CON)

- *Filing/despatching/messaging*: see discussions in Chapter 4 of the economic issues where the end-user is far removed from the OA host or server, thereby imposing, time, distance and cost restrictions or barriers – this is a basic IM issue which in many cases must be addressed *before* the application placement stage, being inherent in the requirement and functional specification. (REL) (CON)
- *Typing/entering/inputting*: while becoming increasingly automated, from the application perspective, in most cases it is still true that closer is better – all other things being equal, the closer the application is located to the data intended for input (in tier and actual physical terms) the better. When the multi-user service model is employed, the application gravitates towards the tier where all (or the bulk) of the data employed is normally resident. In some cases (especially where long-distance communications is involved) data will be batch-input locally and then uploaded at high transmission speed to a higher tier, or vice versa. (REL) (CON)
- *Computing/calculating/logically processing*: see Analysing, above. (REL) (CON)
- *Telecommunicating*: in general, the time, cost and ease of communicating are reduced with tier proximity and physical proximity. (REL) (CON)
- *Meeting*: there is increasing use of advanced technology to support video conferencing and workspace sharing, as well as actual co-functioning, where two or more end-users perform a given task (WI) on one or more pieces of shared information, usually by means of an actual shared (virtual) workspace – indeed, such use of a system is a direct substitute for a face-to-face meeting in many cases – here again, bringing the application, the data and the users as close as possible in tier terms is usually desirable. (REL) (CON)
- *Composing*: this is often an interactive (and self-corrective) activity and system response time is therefore an issue – the application and data almost certainly should reside on the same tier. (ABS)

Beyond the above discussion of work function types, and their impact on the APM, there are various quality assurance (QA) issues not directly considered here. For example, a mainframe may offer a very feature-rich word processing environment, but also impose excessive response times due to long-distance data communication overheads. This is particularly true where the somewhat 'bunchy' X.25 protocol is utilized, and it may more than offset the feature-richness from an overall word processing quality standpoint. It may annoy and impede users who cannot function at their maximum speed and may actually induce some errors. This will be particularly true for users with very high keyboard input speeds. The lag or 'sluggishness' introduced by network delays and mainframe front-end handling may frustrate such users and/or add errors.

2.7 While the purest manifestations of the WFM (addressing function, subject, workplace and time) would, in theory, accord each and every workplace (as defined in the WFM itself) *all* of its respective required TS resources, it is known that this is not economically or operationally practical, at least not for the foreseeable future. Putting ten times the power of your current mainframe on everyone's desk would, in theory, eliminate any concern about tier placement and create a truly one-tier system. However, issues of data correspondence and update/version control would still persist. Further, some classes of WIs are batch-oriented (or are otherwise automatic) such that it may be necessary in the model to consider large systems not only as supporting the workplace but also as themselves being workplaces, at least to an extent. Nonetheless, most WIs still touch workplaces inhabited by real humans. In general, the *degree of workplace-specificity exhibited by WIs* encompassed in the application will determine the degree to which the application can be characterized as 'grass roots' or 'user-oriented'. (REL) (DOWN)

Consider two examples of applications, each with a total WI count of 1000, and each consisting of a total of 150 different WI classes, but with the top ten WI classes in each application accounting for more than 75% of the all-class-inclusive WI total:

- Application A has 600 of its WIs visit only synthetic (i.e. higher tier) workplaces while only 400 are generated by and/or visit the workplaces of one or more of the 32 (human) end-users in the workgroup.
- Application B has 900 of its WIs flow through one or more of the same 32 workplaces.

The more automated WI generation and processing, and the less human involvement there is with this process, the less critical it is to locate the application close to the workers themselves. Increased WI automation and decreased actual human involvement will tend to cause the application to gravitate up-tier. Conversely, increased gravitation of the application towards the workplace level points to an increased need to locate the application down-tier. (REL)

2.8 The degree of *security* required by the unit or workgroup, and reflected in the information actually managed by the application, may impact choice of processing tier. Of course, you cannot determine how much security you need until you have an acceptable and agreed definition of the threat(s) to which you expect the user group and the application to be subjected. Security requirements exert different tier-directional pressures depending upon the degree of security required and possibly also considering the characteristics of the end-user community. Consider the following examples.

(1) In general, and for a contiguously located workgroup or unit, moving the application up-tier increases security beyond the lowest common denominator (basic) level, although the greatest increases in security are exhibited in the move from WS to MBS, from MBS to MRS and from MRS to midframe. (REL) (UP)

(2) As a special case, a geographically diverse (even diasporatic) workgroup will encounter difficulties in seeking to keep an application secure, above the basic level, unless it resorts to draconian measures or else moves the application up-tier to a common service tier. This usually means a midframe or mainframe which is operated by IT (and not by the end-user organization) and where professional database administration and security management capabilities are available. (ABS) (XWS, XMBS, XMRS)

(3) A single-site, and reasonably sophisticated, workgroup can achieve a satisfactory level of enhanced security by using a senior tier system (midframe or mainframe) as in either of the above two cases, but can further increase security to a very high level, if required, by dedicating and even electromagnetically isolating a single-user system, a group of single-user systems or a dedicated multi-user system with shielded LAN or star-wire connections. (REL) (DOWN)

The workgroup/application security requirement thus exhibits unstable behaviour as a tier placement driver. If it is assumed that the entire corporate backbone network (and attached processors) will reach some basic level of security – such as NCSC Level of Trust C2 – this would then become the *default*. Workgroups with this requirement would be tier-neutral in their security-driven application placement preferences because all tiers would provide the same degree of protection. However, even in this (idyllic) situation, requirements above the basic level (i.e. for enhanced security) would still tend to drive the application up-tier. But, as seen above, very high requirements could sometimes drive it down-tier again, and into an isolated (tempested) facility operated by the workgroup. Tempested systems are electromagnetically protected.

2.9 If with new applications (or on an *ad hoc* basis once a comprehensive database administration service and corporate data dictionary are in place), it is desired to exploit the *smart work item* (SWI) *model* set out in the WFM, such a strategy will probably increase system resource requirements, tending to push the application up-tier, at least to an MRS platform. (REL) (UP)

2.10 Where WIs exhibit a high degree of interrelationship, interoperation and interdependence with WIs existing wholly (or mostly) outside the unit or workgroup, the tier-residency or 'tier-gravitization' of such *external* WIs is important in application placement. This may be true if this is a significant factor when a WI census is made on a class-weighted case as discussed in the Application A and B examples in item 7 above. Usually, if a large number of WIs (which each embrace process and data) have external links, then so too will the entire application. Convergence of WIs within the application (and hence the application itself) with the tier or locus of the external WIs is desirable, all other things being equal. (REL) (CON)

System architecture

It is recognized that, in considering factors that drive application placement, the major issues of system architecture and the application itself begin to converge.

3.1 It should be the practice of your IT organization to ensure that a dedicated system (or that portion of a multi-purpose system which is dedicated to the application) possesses a *minimum 30–40% capacity surplus*, in terms of the TS rating, versus the declared TR rating of the application. In some cases, such as where the intended target is a large MRS, this will drive the application up-tier to a larger platform. In other cases, the proposed or existing platform can be upgraded to provide this surplus capacity. In no case will the search for more capacity drive the application down-tier. (REL) (UP)

3.2 It is generally desirable to place the application as *close to the user* as possible. (REL) (DOWN)

3.3 Where the default application is a *workgroup* application, this will tend to drive the application up-tier (at least out of the PC and WS platforms, since they are only single-user), but this only holds true where the application might otherwise have resided at the workplace tier. (ABS) (XPC, XWS)

3.4 Your organization probably seeks to *manage information as a corporate resource*. (If not, it is falling seriously behind the times!) Where information cannot be managed as a corporate resource (to the satisfaction of IT) on the intended tier of application placement, the remedy will almost always be up-tier migration of the application. (REL) (UP)

3.5 Where a software registry has been established, it will be desirable to promote the sourcing of applications there wherever possible. Where an application, thus obtained, is not already running on all tiers, it is possible that such sourcing may connote a tier bias in favour of the tier(s) on which such an application already runs. It could be called a '*home tier bias*'. This is particularly true if the unit or workgroup has a limited budget and the registry itself is unwilling or unable to fund inter-tier porting of such application. (ABS) (Biased in favour of tier residency of application as provided by registry)

3.6 Default security (where it is at NCSC Level C2 or below) can be considered to be the 'norm'. Such a security requirement should be tier-neutral, once the requisite human, physical and system elements are in place at all tiers. Clearly, the former two may be the greatest challenges. Until such security is relatively uniform across the four tiers, increased security requirements (over whatever is the default or norm) will tend to drive applications up-tier until the required level of security is achieved. Otherwise, refer to para. 2.8 above.

3.7 As a default, you can assume that the application has *no requirement for remote access*. In virtually all cases where this condition is violated and such a requirement does exist, it will force the application out of the workplace tier, even if it would otherwise be resident there. The tendency of end-users to shut down their own systems and the fact that PCs and workstations usually offer the lowest physical security and the lower TS rating of such systems all militate against admission of remote users to an application resident on such equipment. Note that DOS-only systems are even less capable of admitting such remote access without complete disruption of (or preferably the complete absence of) the primary user. Note also that this factor does not apply to vertically integrated applications.

3.8 Where there is a requirement for *transaction processing* (which is also treated as a non-default item), there is a two-stage preferential construction:

- For various technical reasons, this requirement absolutely excludes the WS and MBS options in most cases. (ABS) (XWS, XMBS)
- Otherwise (i.e. at the MRS level and above), this need will tend to drive the application up-product line (to a larger model within, for example, an MRS product family) or else simply up-tier. (REL) (UP)

3.9 Where the default of no *parallel processing* is violated, the same logical construction will apply as for transaction processing, except that the absolute exclusion is restricted to WS. (ABS) (XWS), otherwise (REL) (UP)

3.10 Where *distributed processing* is required this will, from any given tier, tend to force the application down-tier. From an architectural perspective, processing becomes more distributed by definition as the application is both moved down-tier and placed on multiple systems. (REL) (DOWN)

3.11 Regarding *network service requirements* (i.e. CR ratings and their detailed service component breakdowns – which are not fully addressed in the SFM) it is possible to postulate a multi-dimensional sub-model. Where we consider WI (A1), the first work item class handled by the application, it will be seen that there are four quantifiable attributes:

- number of WIs in class (usually quoted on an annual basis)
- complexity of WI execution logic that touches the network
- nature and volume of data to be transmitted over the network
- intensity of network transactions:
  - number per minute
  - duration of each
  - origin–destination distances

Together, these would produce an applications 'communications envelope', the four-dimensional content of which is analogous to the volume of a three-dimensional object or physical model. This would permit differentiation of applications which run, for example, on an MRS and which have identical network *connection* requirement but which also exhibit vastly different actual *consumption* of network resources. The treatment of CS resources as 'utility' resources in the SFM is reasonable and it can indeed be extended. To extend the analogy used in the SFM, two users may both be connected to 120 V a.c. electrical outlets, but one may consume far more power (kWh) than the other over a one-year period.

Further, the extension of this sub-model permits consideration of the amalgamated envelopes of all WIs included in the application, here WI (A1. . AN), such that the sum of the thus-calculated 'volumes' represents a characterization of the total 'information displacement' of the application. *Ceteris paribus*, the greater this information displacement, the greater the need either to:

- relocate some WIs (process and data) or at least some data; or
- move to a vertically integrated application.

Such a sub-model would permit exploration of the network impacts of data or data/process relocation. In general, when the logic or process components are themselves distributable over the IM tiers, data relocation to achieve convergence, even where feasible, does not yield greatly reduced information displacement. In other words, when the basic processes (process elements of WIs) are, in their natural or inherent state, or at least in their current state, scattered over the four tiers it is likely that the proposed application's processes will also be scattered. In such circumstances moving the data to fewer tiers without relocation (more difficult to move) processes will not greatly reduce network activity associated with the application. More processes will simply have to go inter-tier to obtain and/or deposit their data.

3.12 The various pure *technology* factors set out above would impact placement as follows:

- CPU, *memory* and *storage* requirements tend to move the application up-tier as they increase. (REL) (UP)
- RDBMS requirements, particularly where interoperability is concerned, tend to push the application out of the workplace tier and otherwise, where the requirements are intense, up-tier. (ABS) (XWS), otherwise (REL) (UP)

This is discussed in more detail in the section entitled 'The application', below.

3.13 Regarding the *client-server model*, placement is:

- most efficient (in the strictest sense) when the server and client are on the same tier (i.e. one MRS is the client and another is the server)
- less efficient when the server and the client are located on contiguous tiers (e.g. MRS is server and WS or PC is client)

- least efficient where the server and client or located on non-contiguous tiers (e.g. mainframe is server and PC is client)

Here, 'efficiency' refers to user interactivity and response times, network resource consumption and cost and also to inter-tier TS trade-off potential at the planning stage. This is not 'user interface efficiency'.

Development and system integration process

4.1 *User desires* will not always be consistent with the principles of good system planning, implementation and management. These may impose ABS or REL impacts and must be assessed on a case-by-case basis.

For example, one workgroup may insist on remaining with the client-server service model and may thus reject any multi-user option out of hand. If their intended application does not technically require a midframe or mainframe scale machine, and is not otherwise migrated there as the APM is employed, their preference is therefore expressible as follows: (ABS) (XWS, plus !MBS or !MRS).

In another case, a workgroup may insist on keeping so much information on-line (or on running a second copy of so large an application in hot-standby mode at all times) that they thereby force their application up-tier, for example from MRS to midframe. Here the preferences are expressed by the notation: (REL) (UP).

4.2 Using a 'continuum of portability' it is possible to assess the *degree of difficulty to port* from a source system to a destination system. The continuum sets binary compatibility at one end and a total rewrite at the other. In many cases a former application will be redeveloped into a new one or at least into a prototype for the new application. In some cases, it may be possible to isolate tier-specific barriers to porting of a given application to, or from, one or more tiers. For example, these can include language, package or operating system versions which differ among tiers (e.g. ORACLE 5 is running on mainframe but ORACLE 6 or 7 is running on MRS). ORACLE, like most quality software products, offers excellent upward (forward) portability, but in this example an application migrating from the MRS to the mainframe might encounter difficulties, due to a backwards port. Where these barriers exist at the target tier, they would tend to foster REL or ABS migration away from such tier.

4.3 At the *Preliminary Classification* stage, a decision to treat either PROGRAM or PLATFORM deployment as the default may eliminate some tiers or at least reduce the likelihood of their becoming the target. For example, dedication of an MRS as a platform at the enterprise level is far more expensive than installing the same system within a workgroup as a general-purpose (PROGRAM) system. At the corporate data centre, the MRS will consume valuable raised floor space in a fully conditioned environment, and will be supported by highly paid IT specialists, not a trained/certified para-professional as in the workgroup setting. Installed within the workgroup, that same system would have a far lower total overhead. Selecting and sizing an MBS or MRS purely as a platform for a single application has three negative impacts, all other things being equal:

- It precludes use of any corporate OA mandating to help fund the system.
- It precludes implementation of other applications on the same system later, if only because the system was sized just for the (current and anticipated future) demands of the one application.
- It virtually guarantees that any substantial growth in application logic, data record size and/or data set size, beyond the planned capacity reserve will either:
  - force addition of a second system, probably splitting the application and/or its data, or
  - force the application up-tier.

This is true of both client-server and multi-user applications. The possible impacts of preliminary application classification can be summarized as follows:

- early selection of PLATFORM (REL) (DOWN) for initial tier selection, but later (if/when the TR demand expands substantially) it will change to (REL) (UP) for subsequent migration
- early selection of PROGRAM (MIN) (usually NEUTRAL, but if there are severe unit/workgroup funding constraints, UP)

4.4 At the *requirements identification* stage there is the choice of *conventional or* WFM *analysis*. This choice may bias the subsequent tier selection in that:

- conventional dataflow and process analysis is insufficiently concerned with the details of 'when' and 'who' as they relate to the end-users with whom, and through whom, the application performs its role – the conventional approach therefore exhibits an upward bias, not being as concerned with the workplace or workgroup locus of activity (REL) (UP)

- WFM analysis may be too concerned with what is happening at the grass roots or workplace level and may thus pull the application's focus too far down towards the bottom tier – it therefore exerts a downward bias. (REL) (DOWN)

There is no way, at this point, to assess objectively the relative strengths or intensities of the respective biases in the two alternative approaches. The organization may correctly wish to combine them into a single (presumably hybrid) approach in an attempt to neutralize (or at least reduce) these countervailing tendencies. In the author's view, no case tool or SDLC effectively combines conventional data/process analysis and workflow analysis.

4.5 The *investment objectives* determination stage is discussed under the *workgroup* section above.

4.6 At the *feasibility* stage it will first be necessary to consider the '*degree of combinability*' exhibited by the candidate applications with the various tiers. If applications from three of the four generic sources are available for all tiers but, for example, the various THIRD-PARTY application candidates can run only at the workplace tier (e.g. they all run only under MS-DOS) this particular generic application source has an ABS tier bias while all others are tier-neutral.

There are limited degrees of freedom in considering combinability because:

- the MODIFY option can include a change of tier as part of the modification, although some tier changes may connote a higher level of effort than others (e.g. mainframe to MRS is probably easier than mainframe to WS or PC in most cases).
- the *ab initio* DEVELOP option by definition can target any tier or combinations of tiers desired.
- the THIRD-PARTY option is symbiotically dependent upon what (if anything) the vendor and the software registry manage to negotiate regarding porting to tiers other than those on which thc application is already resident.
- the REGISTRY option is by definition confined to accepting the application almost precisely as-is, so if it does not already run on a specific tier the choices are to abandon plans to consider that tier or else be prepared to move to the MODIFY option.

There is a range of combinations possible; the variations among them are based primarily upon the following four basic attributes:

- generic source of application (e.g. REGISTRY)
- specific source of application (specific vendor/product name)
- IM tier (e.g. workgroup)
- IT tier (e.g. MRS or MBS)

The maximum number of combinations is represented by the expression:

$$C = T \times S \times M \times O$$

where:

C = maximum number of combinations possible
T = tier (four choices: Enterprise, Area, Unit and Workgroup)
S = generic source (four choices: REGISTRY, MODIFY, THIRD-PARTY, DEVELOP)
M = service model (three choices: single-user, client-server or multi-user)

Assuming no client-server or multi-user service model selection at the workplace level, this renders 56 combinations. Also assumed here is that within the unit (workgroup) tier, multi-user would connote use of an MRS, while client-server could imply use of either an MRS, or an MBS.

The analytical approach itself may determine what is a tier choice driver and what is not. Consider the following alternative methods of analysis.

(1) *Consider generic source first*
This approach finds the best candidate *within* each of the four generic sources, here reducing the maximum number of different real or potential candidates from eight to four, by coupling generic source and specific source together. Then the APM would be used to establish the best tier-pairing (matching application to system) for each of the candidates, and allowing vertically integrated applications (or even two alternative combinations of tier assignments under one candidate) only under special conditions. Finally, the four built-up combinations would be compared based on technical, operational and economic factors and the best combinations, along with its accompanying choice of tier, would be implemented.

(2) *Consider specific source first*
This approach selects the best few applications (or sets the number to be considered at X) without reference to their respective generic sources. Here, X could be set at 4 or any other desired number or the analysis itself could be used to help select the 'top few' contenders. Then, the APM would be used to select the best specific tiers for each candidate with absolutely no restrictions and with nothing treated as a special condition. Finally, the combinations would be compared on the basis of technical, operational and economic issues and the best combination would automatically determine tier placement.

(3) *Consider* all *application issues before system selection*
This approach makes the assumption that all candidate applications, whatever their generic source or specific source, can be ported to (and can actually run on) any desired tier(s). Here, the application would be selected purely on the workgroup's own raw requirements and other non-system-related operational issues as well as on economic issues, but only as the latter relate to the application itself. Once the application selection is made, this choice would be treated as virtually irrevocable. Then, the best tier(s) for application placement would be sought using the APM. Working backwards, the APM would have to address all of those issues not treated earlier including all technical issues, the operational elements of tier/system selection and all economic issues not confined to the application itself.

With Method (1) there may be an inherent bias in favour of the tier invoked by the weakest application. By forcing consideration of *each* generic source, we may also force consideration of a tier which an otherwise very weak application brings with it into the analysis. If this is the only candidate application that invokes this particular tier, perhaps this tier is being given undue and undeserved consideration. In Method (3) the placement is somewhat more application-driven, but this approach could box the analyst into a corner if the selected application has an explicit or implicit ABS preference for, or against, a given tier based on hardware, operating system, package (e.g. database) or other requirements. Method (2) may be the best of the three, but none of the methods is seen to be completely tier-neutral. Therefore, it is reasonable to conclude that the method or process used during at least the feasibility stage of the SDLC – including when and how the APM is to be deployed and used – could itself induce a considerable amount of tier bias. It is not possible to generalize as to whether such biases, where they present themselves as driving factors in the process of application placement, would normally be of the ABS or REL type. Examples of each can readily be contemplated.

4.7 It is also possible to envision an absolute *preference* on the part of the workgroup *for a given generic source*, such as being absolutely in favour of – or against – development from scratch depending upon past experience, perceptions of IT efficiency in overseeing development or other factors. For example, if IT traditionally takes 30 months to deliver an application, the MODIFY and DEVELOP generic sources may be held in quite low repute by the workgroup. Where such a generic source somehow connotes a particular tier (e.g. IT prefers to develop only for the mainframe or for MBS platforms), this also represents an absolute bias against a given tier. (ABS)

The application

5.1 The *service model* (single-user, client-server or multi-user) desired or dictated by the unit or workgroup, or dictated by the workflow and/or application design decisions taken for the MODIFY and DEVELOP options, will usually impact choice of tier. Consider the following:

- Single-user applications expected *never* to be migrated to the multi-user service model will not normally require (nor be permitted to occupy) the mainframe. (ABS) (XELS)
- Client-server applications cannot be run on a purely single-user system. (ABS) (XWS)
- Multi-user applications cannot be run on a purely single-user system and they are highly sub-optimal on an MBS in most cases. (ABS) (XWS, XMBS)

5.2 Perhaps the most obvious, and most powerful, determinant of application tier requirements and placement, at least insofar as up-tier migration is concerned, is the *scale of the application*. In general, as these parameters increase in value, quantity or scale they push the application up-tier. The following are believed to be tier placement drivers:

- *lines of code* (projected or actual) (REL) (UP)
- *complexity of logic* (from functional specification or full design documents) (REL) (UP)
- number of *calls* made to other applications (REL) (UP)
- degree of *interoperation* with applications on other tiers (REL) (CON)
- non-vectorized total *volume of data* from disk input to, and output to disk from, the application (REL) (UP)
- vectorized volume of data *input* to, and *output* from, *other processing tiers* (REL) (CON)
- application-generated actual *machine I/O*, including optical scanning, reading of files, writing to files, keyboard input, printing and EDI input or output (REL) (CON)
- total estimated operating system *processes or instructions* and/or machine instructions (REL) (UP)
- total on-line concurrent *memory* requirements (REL) (UP)

In the most general case, the higher the TR score rating the larger the required system. Note, however, that the current technology permits a machine in the MRS role to have the power of a midframe or small mainframe from the early- or mid-1980s. A large MRS or small midframe of today can offer more power than the largest mainframes of just a few years ago. Both MRS and midframe exhibit price/performance advantages of approximately 15:1 over the previous (higher tier) equipment. Clearly, however, it is not always economic or practical to equip a large MRS or even a midframe with the storage, I/O and support capabilities normally found within a corporate data centre. As the total on-line and near-line storage requirements of a given application escalate they will still tend to force the application upwards, even in a fully balanced and otherwise tier-neutral, four-tier UNIX environment. This is fully consistent with the previously stated intent that the mainframe should ultimately act primarily (although not entirely) as a corporate data server and that most intensive logic processing should occur at the lower tiers. It is, however, important to recognize that some applications cannot run, or at least cannot run optimally, in anything less than a full data centre environment.

5.3 The projected *degree of portability* (see continuum discussed earlier) of an application may be a determinant of whether it should be left on the tier(s) where it can currently run or considered for vertical porting to another tier. While it is obvious that this impacts the cost to implement the application for the unit or workgroup's use the issue does not end there, particularly if the registry is involved. There are in fact three potentially different degrees of portability with which the project team must be concerned:

A source-to-target (to ensure initial application OPERABILITY)
B target-to-registry (to achieve application NORMALIZATION)
C registry-to-next target (to permit application PROMULGATION)

Certainly, from the perspective of the unit or workgroup, Type A is of most concern. However IT must be equally concerned with all three unless it is prepared to preclude the application, at this stage, from consideration for the registry. Here, IT corporate and unit/workgroup local requirements must be balanced or traded off. Perhaps the best application for the unit will be a nightmare for the registry; in such a case the application may simply be exempted from the registry. In other cases, particularly where two essentially equivalent THIRD-PARTY applications are the final candidates, Types B and C (as well as the assistance and corporate site licensing policy the vendor is prepared to offer to IT) may be tie-breakers in both technical and economic terms. The application emerging as winner of this particular comparison may be, in its present manifestation, entirely tier-specific, at least in so far as the unit (which always wants to implement as soon as possible) is concerned. Thus, selecting an application that demonstrates excellent portability characteristics for the future may be a tier-limiting choice in the immediate context. (ABS) (UP or DOWN is context-dependent)

5.4 The 'family history' or *pedigree of the application* in terms of specific source (vendor), tier/platform history, links or ties to languages or packages, operating system/operating environment and various other factors may be important in determining choice of tier. Formerly, packages such as ADABAS, NATURAL, SAS and others were tier-limiting factors, usually invoking the mainframe or midframe class of system. While this is now far less often the case, there are still circumstances where the requirement for a package, a specific version of a package or for specific package functionality may dictate a choice of tier. (ABS) (UP or DOWN is context-dependent)

5.5 The intended *future of the application* can impact the choice of tier for the immediate placement. Where such a future use connotes a change of application scale (as above defined), role and/or user community nature/size, it may be desirable (or even mandatory) to make provision for these changes at the outset, even in the version supplied to the initial user community. In these circumstances, such provision may in turn determine the tier or at least push the application up-tier or down-tier. The provision could touch not only scale but also such areas as package-dependencies, I/O, transaction processing, real-time requirements, security and other requirements. (REL or ABS) (UP or DOWN is context-dependent)

5.6 Application *support requirements* are an important determining factor of tier, since each tier by definition has unique support capabilities. In general the following are the support characteristics of the tiers and specific component systems:

- Workplace: self-service by end-user
- Unit: *MBS* – para-professional LAN administrator covering PC connectivity, LAN, WAN access, server and client portions of application and any packages such as client/server OA if available – may also include application and/or database administration and OA administration

  MRS – para-professional MRS administrator covering PC, PC connectivity, LAN, WAN access and usage, multi-user application(s) running on MRS and any packages such as OA – usually also includes application and/or database administration and OA administration
- Midframe/mainframe: comprehensive professional support by IT unit, including support and administration of all hardware, software and communications environments

The above list is not comprehensive, although it should provide enough information to assist your organization to decide what the important placement determinants are. Note that it is possible to take a 'common law' or a 'codified' approach to placement. Certainly, the latter is preferable if you can devise a sufficiently flexible (and perhaps automatable) method.

Generally, a requirement of the part of the application itself and/or the users of the application for enhanced service (over any defined default level of support) has the effect of pushing the application up-tier. (REL) (UP)

Economic issues

6.1 The ratio of application-dedicated technology to the total amount of technology to be made available to the unit or workgroup indicates the relative importance of the application in terms of the unit or workgroup's technology base. Where this can be ascertained, it is the *ratio* of TS(A)/TS(T) where TS(A) is the TS value dedicated to the application and TS(T) represents that total capability of the system. *Ceteris paribus*, the following factors will apply:

- the higher this ratio, the more important it will be to the unit or workgroup to have a reliable (in some cases even a redundant) system (REL) (UP)
- the higher this ratio, the more the unit or workgroup may seek to have a system which is closer to its own control, perhaps even fully dedicated (REL) (DOWN)

Thus, these two forces or factors are actually working against each other; they are countervailing in technological terms, but the resolution of any conflict between them will often be based on economic grounds.

There is also at least one special case. The support program for workgroups, established and maintained by IT, may envision both a default and an enhanced level of support being offered to each workgroup. In an enhanced support arrangement, it may be desirable to establish a 'buddy system', wherein paired MRS sites serve as emergency data and operational backups for each other, with special provision for regular reciprocal backups and periodic up-loads of data to a higher-tier systems, with the application also available in a cold- or hot-standby mode on that system. A more drastic alternative is a fully redundant backup (mirror image) system running at the workgroup's site, although this does not protect well against local threats. If the workgroup must pay IT some extra annual fee for such an enhanced level of service, this cost would probably still be less than the annual portion of the fully allocated life cycle cost of an on-site backup system. (ABS) (!MRS)

6.2 The projected fixed and variable costs of LAN and WAN implementation and usage (where charged back to the workgroup's budget) to support communications between the projected application and another application or an OA package are relevant to tier placement. These are *attributable communications costs*. In general, project analysis and intended application user communities will have the following order of preference (from greatest to least) in considering the communications activities of the projected applications:

- same platform
- same LAN, different platforms
- same facility, different LANs and platforms

- different facility, same geographical area
- different geographical area

This preference tends to have the following impacts:

- Where the application and whatever it is to communicate with (another application, an OA package etc.) are both being placed at the same time, *ceteris paribus*, both will be forced down-tier and towards the same platform within a tier. (REL) (DOWN/LAT/CON)
- Where the application is new, but whatever it is to communicate with is not, then this factor will tend to force tier-convergence of the application and the entity with which it will communicate. (REL) (CON)

6.3 Where a THIRD-PARTY application vendor's *product pricing policy* – across platforms of different scale – is not approximately linear it may exert a tier bias. Here we are using a cost/user basis in terms of the maximum number of users licensed by such a vendor for the platform in question. We are assuming that a package licensed for 20 users will actually be used by that number concurrently. This is not often the case in real life, but it gives us (for calculation purposes) identical theoretical and actual cost (also called 'effective' cost) per user. For example, if ABC Ltd offers their accounting package for UNIX systems which you are running now at all four tiers but the effective cost per user is $300 on an MRS and $1500 for a midframe, there is a definite cost bias against up-migration. Such price biases could drive the application either up-tier or down-tier, depending upon the specifics. Further, if the proposed cost of an as-delivered application from the MODIFY or DEVELOP generic sources exhibits the same tendency, such a bias could also impact placement. As a default, it should reasonably be expected that while the size and complexity of the total ready-to-run version of Application X will increase as it moves up-tier, the number of potentially licensable (or otherwise permissible) users should increase even more rapidly, as one moves from MRS to midframe and from there to mainframe. Therefore, all other things being equal, an application (particularly one sold for an open systems environment) should exhibit 'progressive' price behaviour; that is, the cost/user should fall as the application moves up-tier. An application which behaves in the opposite manner exhibits 'regressive' price behaviour.

6.4 The total share of technology, communications and support costs, both capital and operating costs, allocatable to those systems or components dedicated to (or else held in dedicated reserve for) the application must also be considered. While these can be considered in their totality, it will usually be beneficial to compare the *cost per TS*, *cost per CS* and *cost per SS* point in the score ratings of what is actually to be supplied on the various tiers under alternative candidate combinations of application and computer system. This approach can be used for both PROGRAM and PLATFORM application placements. The only significant anomaly may occur where the corporate charges for access to common service equipment (usually mainframe and midframe) which are levied upon user communities either understate or overstate the true cost of providing such facilities. Depending upon the precise system requirements of the application, such cost comparisons (especially where a single candidate application is combined with different tier systems and then competes against itself) may drive the application up-tier or down-tier. (REL) (UP or DOWN)

6.5 There is also the issue of special case pricing policies which may relate to the MODIFY, THIRD-PARTY and DEVELOP generic sources. The software registry may wish to spur promulgation of a particular application, or even the package upon which it is dependent. An independent software vendor (ISV) may offer a very special discount to 'get in the door' of a large customer. Similarly, a contract software developer may offer lower than normal per diems or a fixed price on a substantial project. The project team – and the user community – must not seek out or solicit such bargains on a basis that would bias the tier selection analysis.

There is also another matter which must be considered. The software registry itself must take special care to avoid bias (in working with the project team) as it could be considered, in some circumstances, to be in danger of being in an implicit conflict of interest, as follows:

- the REGISTRY sourcing option brings revenue to the software registry (if the 'used car lot' model is used, wherein software is recycled and sold back out into the user base) and therefore rewards a past decision to include Application X in the inventory.
- the THIRD-PARTY option will in most cases involve the software registry in paying out some incremental premium (over whatever the user group would have paid for a single licence) to obtain a corporate site licence for the incoming application.

This natural (and built-in) desire to promote the REGISTRY generic source option (and failing that, presumably, the MODIFY option) must be recognized by senior IT management, the software registry manager, the project team and the intended user group. There will, of course, be many cases in which the REGISTRY or MODIFY option is indeed the optimum one, all things considered.

Special-case pricing policies may drive the application either up-tier or down-tier during the placement analysis. (REL) (UP or DOWN)

☛ The person running the software 'used car lot' will always prefer that you select from their current inventory . . .

6.6 *Ceteris paribus, implementation and support costs* (apart from the cost of the technology components themselves) will rise with tier due to increased overheads for progressively larger systems. Where all other factors are equal, this would presumably spur the unit or workgroup to seek down-tier movement of the application so as to control costs. Such down-tier movement would stop when the lowest technically and operationally practical tier is reached. (REL) (DOWN)

6.7 There is also the issue of the '*value optimality*' of MRS horizontal *application porting*, versus horizontal application porting on other tiers. In general, there is more to be gained porting applications among MRS platforms than, for example, among PCs or mainframes. PC software investment is lower; mainframes have few horizontal peers. It is probable that MRS-to-MRS porting will yield higher benefits per dollar invested than will either horizontal porting among the tiers or any form of vertical porting. Thus, when the returns to prospective *future* (versus present) porting investments are considered, there may in many cases be a tier bias in favour of MRS. In other words, thinking forward, MRS will be a good place from which to originate the porting of an application in the future. This is because:

- MRS-to-MRS porting probably offers the highest return for the porting money invested, which replicates a (by then) software registry-owned application for a whole new group of users.
- Ports from MRS to larger equipment generally provide access to less restrictive environments, not requiring substantial migration to service the same or a growing group of users.
- Ports from MRS to small equipment will (where the application started out as multi-user) at the very least involve a change of service model for the application and in some cases major changes to logic, data or data handling.

This would appear to be an absolute bias in favour of MRS. (ABS) (!MRS)

Other factors

7.1 *Project team bias* can, of course, have a relative or absolute impact on tier selection. It is usually desirable that these biases be minimized or neutralized. (REL or ABS)

7.2 In certain cases there will be a *directive* from the user community (or their own line management) regarding the selection of platform. This directive may be made for technical, operational and/or economic reasons; however, any such directive is probably subject to functional review by IT.

Application placement framework

This section combines the results of previous tasks which have been addressing architectural, economic and operational issues related to the issue of to which tier (within a multi-tier system architecture) a given application should be targeted.

The mandate of the application placement methodology (APM) is to address:

- all technical issues
- all operational issues that are not application-specific
- all economic issues that are not application-specific

This book does not provide a comprehensive and final version of the APM. For example, depending upon your circumstances, some of the rules presented in Appendix B may require more scientific measures or scales of various input information and/or more rigorous (more quantitative) specification of the rules themselves. In most cases, the improvement of the rule specification will result in an increased demand for input information to that factor/rule. It is believed, however, that the factors included in Appendix B cover all or at least most of the issues that must be considered in placing a given application on one or more tiers.

Key concepts and operatives

1. It is expected that, eventually, a fully automated application planning methodology can employ all of:

    - Workflow methodology (WFM)
    - System functional model (SFM)
    - Economic model of information technology (EMIT)
    - Application placement methodology (APM)

Nothing has been done in the development of any of these components to preclude their integration with the others.

2. There exists a natural order of precedence among issues, which are TECHNICAL, OPERATIONAL and ECONOMIC. If an application, a system or combination of the two is technically not feasible then there is no point considering it. If something is technically feasible but not operationally workable, it is moot. If something is both technically and operationally workable but cannot be justified from the COBA and business case perspectives then it too is moot, at least for the time being.
3. While there can be considerable debate about the choice of service model (among single-user, client-server and multi-user), it is possible to develop a set of rules to assist in the determination of this question. In virtually all cases, this issue must be addressed *before* the APM is utilized.
4. In general, EMIT and the APM must be used together, perhaps iteratively, to achieve the best application placement. Both are used during the feasibility stage of the SDLC.
5. The SFM is viewed primarily as a supply and demand model and as one capable of highlighting hierarchical inconsistencies and constrictions which may be introduced during system buildup or by a decision to utilize or upgrade existing equipment or else to increase the workload and/or number of users for an application. The APM is concerned mostly with the supply side of the SFM; demand is of interest mainly due to its impact on what it is feasible or desirable to supply. The SFM is *not* tier-biased because even the most basic service (TS = 1000) can be provided in any of three 'precise equivalents' which have different tier residencies, specifically:

   - PC286
   - TG/MRS or XT/MRS (graphics terminal or X-terminal with MRS)
   - DWG/MRS (diskless graphics-capable workstation with MRS)

   The model hypothesizes the supply of basic system resources (CPU, memory and storage) plus rudimentary input/output (I/O) capability inherent in CRT and keyboard/keypad. There are three supply modes: absolute (ABS), relative (REL) and replicative (REPL).
6. A default client/server workload distribution of 50/50 is assumed.
7. A manual or automated version of the SFM will be utilized to build up overlays of the triple stack structure (each representing a system architecture component) to fully support each alternative

application/system combination to be considered. Where two or more of these differing application/system combinations contain the same application, there is by definition, an application placement decision to be made.

8. Regarding UNIX, the degree of contigiuity in *time* and *tier* impacts the degree of inter-tier processing power trade-off which is possible in the client-server and multi-user service models.
9. The determination of which of the application driver decision sequencing methods (pp. 145–8) to use must be made at the outset of the feasibility stage or, at least, before the APM is utilized. The determination must be made on the basis of clearly set out rules or decision parameters. These must be as neutral as possible and must have as their objective the downstream maintenance of tier neutrality throughout the tier selection process.
10. It is necessary to construct a default profile for the placement of the application. This is a 'straw man' placement which may well be overturned by subsequent analysis, but it permits the analyst to establish a starting point. By far the most important determinants of the 'default placement' for an application are '*lowest practical tier*' and '*workgroup*'. When these two act together, they usually point to Tier 2 – midrange system (MRS) or micro-based server (MBS) – as the most logical result. Within this tier, the client-server or multi-user nature of the application would finalize system selection.

    It will of course be argued that for either Tier 2 system, but particularly in the case of the more costly MRS, such a default placement strategy is a grossly unreasonable assumption, since most of your workgroups may not already have a new-generation UNIX-based MRS; many may have old minicomputers or just smaller MBS equipment. On the other hand, mainframe or midframe capacity is not free either, even within the UNIX environment. The initial or default placement of the application can be referred to as the default application locus (DAL). The tier of this placement is referred to as TIER(DAL).

    There is also the fact that the service model, and various other attributes, of a given application may vary the default placement right at the outset to some other tier. Thus, the effect of consideration – within the APM – of all placement factors collectively will be to either:

- *confirm* the application placement on the default tier/system
- *displace* the application UP or DOWN, and/or displace it laterally, to another tier/system or else to divide and distribute it among two or more systems within a given tier

Therefore the APM is a displacement model using a sort of vector sum of the displacement effects (or 'votes') of the individual placement factors. Put another way, the 'overlay' of all individual vectors of displacement created by the operation of ABS, REL and MIN factors (impacting the technical, operational and economic areas of concern) results in a 'net vector' which either does or does not displace the application from the DAL. See Fig. 5.1.

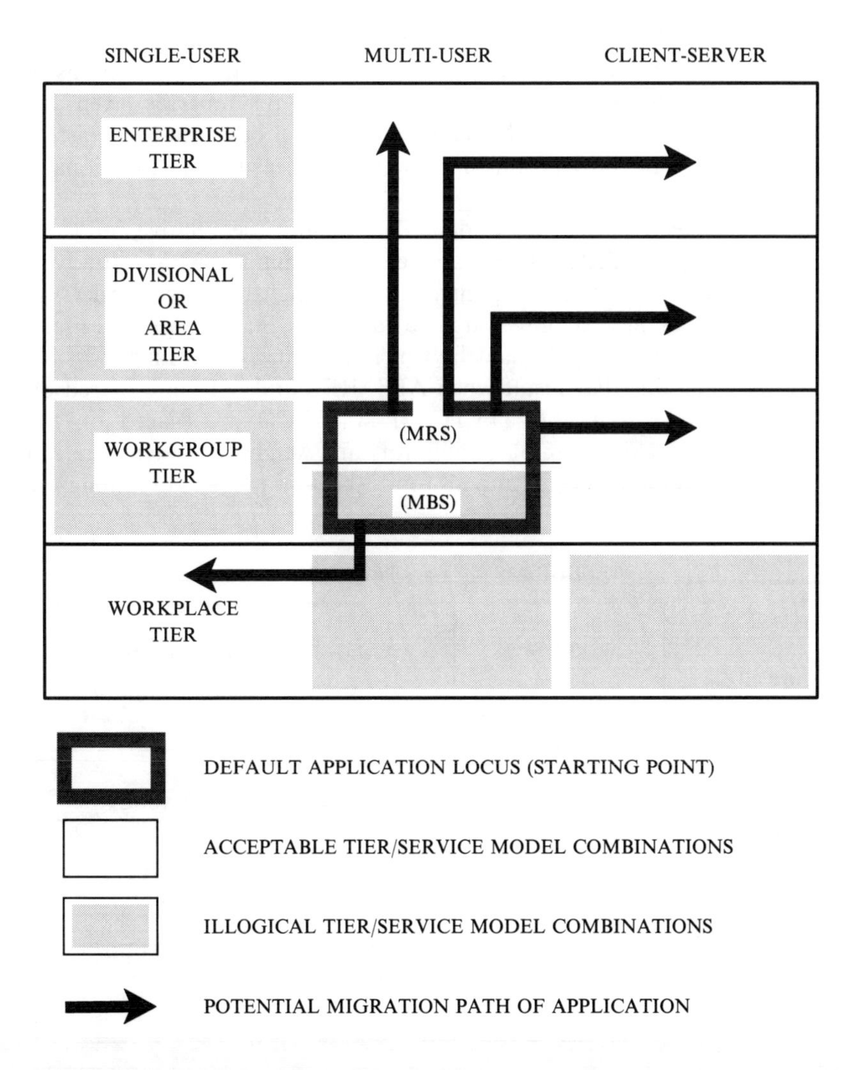

Figure 5.1 Tier displacement model.

11. Not all placement factors should be accorded equal weight or impact in determining tier displacement. For example, it was established above that technical issues must be considered as being prior to operational ones, which are in turn prior to economic issues. Similarly, it is reasonable to assume that an application placement driver (and its resulting factor/rule in the APM) that dictates an ABSOLUTE placement to a given tier should have a greater impact than a RELATIVE factor which merely points in a direction such as up-tier or down-tier. MINIMAL factors should be accorded still less influence. While there may be a possible 'work around' strategy even for an ABSOLUTE factor, the fact that in most conditions it conclusively includes or excludes a specific tier must be given due consideration in the APM. To fail in this respect would be to distort the result, possibly producing technically infeasible or operationally ludicrous results.
12. There is no perfect method of determining the relative influence which a given factor should be accorded within the APM. However, it is believed useful to jointly consider the above two orders of precedence in assigning weight to a factor. Stated more precisely, Fig. 5.2 represents a matrix which recognizes the nine possible combinations that result when ABS, REL and MIN are placed on the vertical axis and TECH, OP and ECO are placed on the horizontal axis. Using a straightforward weighting system, we can now let all left-to-right and all top-to-bottom movement among the

	TECHNICAL	OPERATIONAL	ECONOMIC
ABSOLUTE	VALUE = 9	VALUE = 8	VALUE = 7
RELATIVE	VALUE = 6	VALUE = 5	VALUE = 4
MINIMAL	VALUE = 3	VALUE = 2	VALUE = 1

Figure 5.2 Nine-block matrix for application placement model.

nine blocks result in a weighting decrement of one or more. This is accomplished with a simple matrix with the top-left block valued at 9 and the bottom-right block valued at 1. Of course, this matrix permits ready classification of each of the various application placement drivers into one of the nine blocks.

13. Assuming each factor results in either a null vote or a vectorized displacement vote, these votes can then be resolved against the weighting of their respective blocks. There is the issue of whether to effect such a resolution at the block level or at the whole matrix level.

Block-level resolution offers two advantages:

- vectors are traded off only among like-valued factors
- fewer total displacement scores are produced, so vector traceability is maximized and it is easier to diagnose errors

There are, however, problems with this approach:

- it will magnify possible errors or distortions in high-value blocks
- it aggregates (losing much detail) very early in the overall train of analysis

Matrix-level resolution offers the following advantages:

- More detail survives until the end of the analysis.
- The number of factors present in a block is variable and also helps (subject to weighting) to impart that block's degree of influence on the outcome – thus, for special case placements with many custom factors not only the existence but also the *total context significance* of the custom factors is taken into account – in most cases such case-specific factors will be in the lower weighted blocks (note, however, that the very weighting of these blocks still militates against a user group 'stacking' the analysis).
- Finer resolution and traceability is provided, especially in cases where the net result represents a fraction of a whole tier displacement.
- It is possible to compare directly the value contribution (to vectoring) of *any* two or more factors, without regard to the factors' home blocks.
- A factor which is divided over two or more blocks (has manifestations in each because it cannot be cleanly classified as falling only in one block) can be executed in any order desired without upsetting previous analysis.

The only real disadvantage of matrix level resolution is that more calculations are required and the resulting interpretation is somewhat less intuitive.

14. While the distribution of the application placement drivers found in the this chapter provides a reasonable population in all but the lowest valued blocks it does not provide an equal distribution among blocks. For this reason, and because there may be cases in which unique factors appear, it is desirable to have a method for the introduction of new, non-standard, *custom factors* into a specific analysis. Use of an expert system to support the introduction of custom factors would not only permit an interactive (and guided) session for this purpose but would also:
    - filter out spurious factors or those of negligible import
    - permit the APM, over time, to 'learn' and thereby to effect generalization of new factors, which become less parochial, even mainstream, factors, as the art/science of application placement evolves based – among other things – upon an expanding body of precedent
15. The APM will determine the final (net) displacement caused by all of the stock and custom factors considered, where each of these is executed in the form of a rule, with the appropriate weightings assigned to the results. In some cases, interpretation of the overall result will be required. The result should be the selected or optimum tier, given the factors considered and the information utilized within each factor.

User taxonomy

1. User groups can be classified by such items as:
    - degree of sophistication
    - current system environment(s)
    - role or function in relation to the application itself
    - resource requirements (TR, CR and SR, as in Chapter 3).
2. In setting a unit or workgroup's user demand profile (UDP), per Chapter 2, it is necessary to use a THRESHOLD Approach. Each user must be accounted at his or her most demanding use. A user who uses simple 3GL application programs (dependent only on the OS) with a demand load of TR = 1000, but who also uses an application dependent upon ORACLE, must be counted at TR = 3000, not TR = 1000.
3. Of course, the SFM point scores generalize user requirements into a one-dimensional measure (or, more precisely, into three-one dimensional measures). The introduction of additional quantitative

and qualitative information will be of assistance in understanding and interpreting the true requirements of the user group.

4. Provided IT project teams are unbiased (regarding tier and technology selection issues), and provided their approach to user groups is consistent, the case-specific or custom placement factors introduced by a given user group provide a good window into the unique aspects of that group. In a broader sense, across many user groups over a period of time, the nature and thrust of such factors permits an additional taxonomy of units or workgroups. As stated above, an expert system 'filter' for the entry of custom factors would not only prevent spurious or duplicative factors but also permit the APM to 'learn' more about the total universe of potential custom factors, as it is revealed through successive application placements in different projects.
5. To a certain degree (and the issue is not fully addressed in this book), there is a symbiotic relationship between various attributes of the unit or workgroup and the choice of application service model. For example, a workgroup of certified professionals in a technical discipline will probably demand a greater portion of the application logic be executed close to them (and hence a higher degree of desktop machine intelligence) than a group of data entry clerks.

Application taxonomy

1. There are a number of ways in which an application can be classified. These include:

   - service model
   - scale
   - default application locus (DAL)
   - dependencies and requirements
   - generic source
   - specific source
   - past, present and future manifestations

   These means of classification provide an external cross-check or yardstick of the overall reasonableness of a given result produced by exercising the APM. They apply known means of 'sizing up the beast' in relation to the recommended tier placement. By analogy, the coefficient of drag or the thrust-to-weight ratio of an aircraft designed with extensive use of computerized models can still be considered manually by the aeronautical engineer. The engineer would, for example, *not* expect a transport aircraft to exhibit a thrust-to-weight ratio of 1:1; if it did, he or she would immediately suspect a verification or validation problem with the model, or at least faulty input!

2. The choice of service model tells us a great deal about how the application will perform its share of various automated and semi-automated work items, how it will relate to its data and how end-users will interact with it. It provides information both about what tiers cannot be employed (workplace is banned for all but the single-user model) and about what will happen if too many tiers separate the application and its data.
3. The scale of the application, as characterized above, provides a collection of measures which in general provide information about the application's vector tendency; i.e. it is either neutral, tends up-tier or tends down-tier. In certain cases the scale of the application will, by itself or in concert with very few other factors, determine the placement of the application. For example, an application certain to have more than 2000 concurrent users would in most cases be elevated to the enterprise tier, either to a mainframe or a dedicated supercomputer.
4. The default application locus (DAL), discussed in the next section, provides important information about the *preliminary assumptions* or the *apparent nature* of the application. These early findings may be either confirmed or overturned by subsequent analysis in the APM; however, there is nothing wrong with adopting them at the outset. In all cases, the DAL can be used to classify applications provisionally as to their prime tier of focus.
5. Application requirements and dependencies, such as the need for a given RDBMS or the need to interface with another application or a manual system in a certain way, also permit classification of applications. A review of the application placement drivers will indicate a number of such relationships, mostly as related to system architecture and applications.
6. The generic source of an application (i.e. whether it comes from the REGISTRY, MODIFY, THIRD PARTY or DEVELOP source) tells us a considerable amount about its history and cost, and in many cases provides information about its tier placeability or flexibility in various circumstances.
7. The specific source of an application (i.e. the application name and supplier name, where applicable) also imparts information about the placeability of the application, although this is much more context-dependent and less reliable than the more general assumptions (and hence easier classification) possible under generic source.

8. The past history, intended use after application placement and the projected future manifestations of the application provide important information to assist in application classification process.

Application placement methodology

Overview

The APM, while narrowly defined above as a matrix-driven displacement model, both impacts and is impacted by several other factors. Thus, a wider view (encompassing much of the feasibility phase of the SDLC) must be taken. In this context, the following activities relate directly to the placement of the application:

- determination of the 'feasibility sequence', particularly in what order to consider generic sources and specific sources of the application, and hence where to use the APM itself
- selection of application service model
- establishing the initial 'straw man' placement of the application (called the default application locus)
- determining whether the entry of custom factors is required
- collection of all necessary information for stock and custom factors
- invoking the APM, thereby generating, weighting and interpreting application displacement votes both individually and collectively
- applying the net displacement to the default application locus so as to reposition the application if required
- cross-checking the resulting recommended placement

This section provides a more detailed discussion of these steps and is divided into appropriate sub-sections for this purpose. Note that cross-checks are dealt with in the next section.

Step 1: Feasibility sequence method selection

1.1 As was pointed out above, the choice of whether to consider generic source first, specific source first or to consider all application issues before placement can influence the outcome of the APM. The following rules will assist in the determination of which method to use; this determination should be made at the outset of the feasibility stage and in all cases before the APM itself is utilized. While a TRUE response to even one rule should be sufficient to invoke the respective sequencing choice, in most cases two or more such rules will apply. The following method should assist in resolving any conflicts:

(1) Signify as TRUE all valid rules.
(2) If no methods have TRUE rule conditions select Method A.
(3) If only one method has TRUE rule(s) select that method.
(4) If two or more methods have TRUE rules then total the numbers of the TRUE rules under each method and select the method with the lowest score (the rules are ranked in importance from greatest to least).
(5) If Step 4 produces a tie select Method A.

1.2 Method A is justified as the default because we want to encourage consideration of the REGISTRY and MODIFY generic source options as much as possible so as to maximize the return to the organization from the portability and scalability advantages of open computing systems where potential for same exists.

Figure 5.3 provides a schematic of the operation of the three sequencing methods.

1.3 Rules for Method A: GENERIC SOURCE FIRST

A1 This is the *default* and is to be used if no B or C method rules apply or else to break weighting score ties between A, B and C rules. This rule should also be invoked if Rules A2 or A3 fire. Rule A2 has precedence over all rules except this one.

A2 There are potential *candidates for 3 of 4* GENERIC SOURCES.

A3 The unit or *workgroup is replicative* and your organization is not believed able to negotiate a corporate site licence under one or more of the possible candidate applications under the THIRD-PARTY option.

1.4 Rules for Method B: SPECIFIC SOURCE FIRST

B1 There is a *prima facie* case that *each application candidate* is completely *tier-specific* and the tiers vary across the candidates.

B2 There are absolutely *no registry applications* relative to this requirement, even when the MODIFY generic source is considered.

B3 The choice is narrowed at the outset to *make or buy* and the make (DEVELOP) or buy (THIRD-PARTY) options each have only one candidate seen as being a viable contender from each of these two generic sources.

B4 The unit or *workgroup refuses* to consider two or more of the GENERIC SOURCES and IT gives its functional approval to this refusal.

B5 *Economic constraints* preclude consideration of the DEVELOP or THIRD-PARTY options.

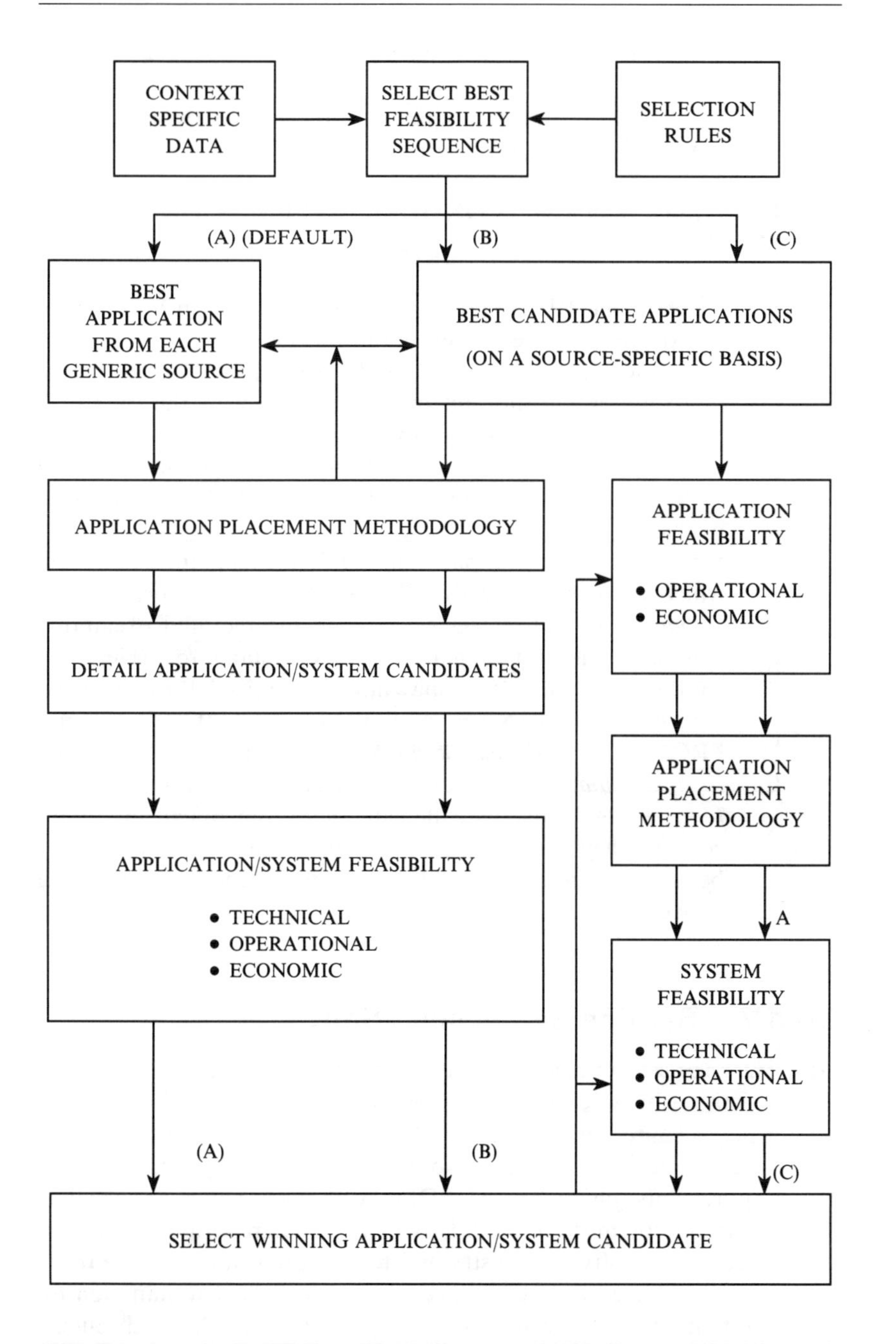

Figure 5.3 Alternative feasibility phase sequences

B6 *only* THIRD-PARTY application candidates are considered and all of these are already in service within the organization, but they have not been entered into the registry yet (i.e. you have *no corporate site licence* for any of them).

1.5 Rules for Method C: CONSIDER APPLICATION BEFORE SYSTEM

C1 At least *one* REGISTRY *application* could clearly be run in any of the top three, or even all four, tiers.

C2 Tier placement and service model are *pre-decided* by one or more non-application-related issues known at the outset. A *prima facie* case therefore dictates tier. Thus, all considered application candidates will be paired with that tier only and with that service model only. (The APM is effectively bypassed.)

C3 Workgroup and/or application requirements point to a system which *violates* the UNIX, POSIX, OSI or other *open system standards* and IT agrees to provide the required exemption from corporate functional guidelines or standards. (Here, the choice of system may or may not dictate tier but often such a choice will point to a PLATFORM placement rather than a PROGRAM placement when it comes to the actual machine.)

C4 *Each candidate* application is practical only on a *subset of the four tiers* and, together, the candidates' tier practicality maps conflict; they are not all practical on the same tiers. (This may be another case of the APM bypass instituted by Rule C2 above.)

Step 2: Service model selection

2.1 The rules set out below, derived from the above material, will be of assistance in selecting the service model for an application, where generic source or specific source candidate selection has not determined this already. Note also that in many cases the requirements phase of the SDLC will have made an absolute determination in this regard. However, if you are serious in wanting to develop a software registry and reuse workgroup software (once developed or corporate site licensed) such early determination (at the requirements stage) should be made only where absolutely necessary, or else many otherwise viable candidates may be eliminated very early in the SDLC.

2.2 The rules set out below are strictly related to the service models themselves, with minimal consideration of tier issues. This stringency has been required because, except at the workplace tier and the MBS element of the workgroup tier, all three service models are possible at all tiers. Not all are optional, however.

2.3 No specific order of precedence or tie-breaking method has been included for the rules set out below. Note, however, that it is desirable to consider 'workgroup' and 'multi-user' as default characteristics of the application (as in the DAL determination below) unless it can be established otherwise. These will maximize future portability.

2.4 Rules for single-user model

S1 The application has only a *single intended end-user*, or else one end-user access at a time per installation or image of the application, *and* there are no plans for any other type of end-user access to be made.

S2 All of the *processes* and *data* inherent in all (or at least most) of the work items (WIs) included in the application exist on a *single tier* and are utilized by and/or involve a single end-user.

S3 The application was expressly developed for, and is intended to be run under, a *language confined* to a single-user operating system (or operating system image) such as MS-DOS, OS/2 or Mac OS.

S4 The application was expressly developed for, and is intended to be run concurrently with, a *package confined* to a single-user operating system.

2.5 Rules for client-server model

C1 By definition, the application requires some degree of common processing and/or common data, and also, separately, requires the action of worker- or workplace-specific processes or data created and held by an individual. A significant portion of the *common data and individual data* is exchanged or is at least acted upon by the other party in the client-server relationship.

C2 The *distribution of process execution* between the workgroup and individual work environments is *unknown* at the outset, but for purposes of this application the individual can be shown to have an absolute requirement for some degree of data processing and/or storage capability to be delivered at the workplace tier. Further, the potential for a vertically integrated multi-user application either does not exist or is restricted in this case.

C3 There is a requirement for workgroup data to be held in common, but there is no requirement for two or more users to share a database file (or record) under any circumstances. There is, in other words, *no concurrent shared access* (whether restricted to shared viewing or extended to jointly or in parallel applying processes) to common data.

C4 The requirement or opportunity exists to *share I/O devices* necessary to the application and this requirement *cannot* be otherwise accommodated on the same tier under the single-user model. (Note: caution must be exercised here if the single-user application is, for example, already intended to be run on an MRS for some other reason but is still made available only in single-user mode. In such case, this rule *cannot* be invoked to effect migration to client-server. Specifically, the single end-user's UNIX application could share all MRS peripherals with all other applications and processes running under UNIX on the system. In other words, this requirement alone is insufficient to transit the workgroup to the multi-user model.)

C5 The unit or workgroup will implement *client-server OA*

C6 At least 30%, and preferably 50–80%, of the total application *workload can be downloaded* from the server to the served clients. This percentage holds true throughout all normal production operation, backups, restores and contingency/emergency situations.

C7 When the server is already known (or intended) to be an MRS or MBS, use of the client-server service model with a LAN *does not violate application or workgroup security* requirements. This includes the requirement to safeguard the existing level of security on such system. Where the system is an existing MRS already running multi-user applications the security of these applications too must be safeguarded during the choice of service model.

C8 There is a significant *benefit* to the *redistribution of data* and/or logic in a previous or parent application into the client-server model.

2.6 Rules for multi-user model

M1 *Two or more users* will regularly, and concurrently, invoke all or a portion of the application to:

- perform the same process on different data
- perform different processes on the same data
- attempt to perform the same process on the same data

but must be prevented from doing so or else permitted to do so only under restricted (and defined) circumstances

M2 The application requires *specific RDBMS capabilities* normally associated with a multi-user environment.

M3 The application runs at the MRS tier and/or higher tiers and under no conditions can download of application data to the workplace tier (PC or WS), nor upload of data from the workplace tier, nor data storage at that tier be permitted to disrupt or halt network and/or workplace system operation. This is the '*workplace critical*' rule.

M4 *No* significant *benefit* can be shown from *duplication* of all or part of the application logic and/or data on two or more processing tiers or processing platforms. This being the case, it is understood that such duplication incurs at least operational and economic costs (and sometimes has technical implications too), but provides insufficient benefit.

M5 The application and/or end-user invokes considerable *background processes* during normal operation. In a client-server environment these processes might consume both network and system resources (incurring both CR and TR), but in a multi-user scenario they would normally consume only system resources.

M6 The application is intended to be *vertically integrated* with logic resident on, and to impact data normally resident on, two or more tiers but *without significant interchange of data* among such tiers.

M7 The unit or workgroup, with or without resorting to any future corporate mandating of OA to all information workers, intends to implement OA concurrently with the advent of the application and there is a requirement for *multi-user OA*.

M8 There is a requirement for significant interworking of the OA and RDBMS packages in support of the application *and* one or both of these packages is required to support (or by its own nature itself requires) a full multi-user model.

M9 A given RDBMS data set or file is to be *accessed* simultaneously *by two or more applications* residing on the same processor, or at least the same tier.

M10 There is a *security* requirement, known at the outset, which exceeds the capabilities of any client-server system certified by IT for use in the organization and any such system able to be readily so certified.

Step 3: Establishing the default application locus (DAL)

3.1 The DAL is established by re-setting any of the *default indicators* set out below according to what is known about the application before the APM factors/rules are exercised. In many but not all cases, re-setting a default indicator will cause a change of default tier.

The indicators will allow a preliminary determination of the value of TIER(DAL), where enterprise is valued at 4 and workplace is valued at 1.

The following are the default indicators.

D1 The *deployment strategy* will be first to prototype and then to production.

D2 The WI(A)/WI(T) *ratio* will be less than or equal to 50%.

D3 WI *flow* will be at least 60% intra-workgroup.

D4 The acquisition model will be PROGRAM, or the target system will have been acquired previously under the same model.

D5 Per Maxim 1, the application will be placed on the *lowest practical tier*.

D6 The operating system environment will be UNIX + POSIX.

D7 The GUI will be your default *standard GUI*, if feasible.

D8 LAN and WAN communications will *conform* to OSI standards or to TCP/IP.

D9 The application will *conform to other key open system standards* which do not permit anything less than a true multi-user application.

D10 The operating system and hardware environment will *conform to a standard profile* which your organization has established.

D11 If an RDBMS is required it will be Product X (your choice of RDMBS). (You may vary this according to your organization standard.)

D12 If an OA package is required it will be Product Y. (You may insert your corporately mandated OA package, if such exists, here.)

D13 There will be *no remote access* requirement. (If this rule is violated, set TIER(DAL) = XWS.)

D14 There will be *no transaction processing environment* requirement. (If this rule is violated, set TIER(DAL) = XWS, XMBS.)

D15 There will be *no parallel processing* requirement. (If this rule is violated, set TIER(DAL) = XWS, XMBS.)

D16 This is *not a vertically integrated application.*
D17 There will be *no distributed processing* requirement.
D18 This is a *workgroup application,* not a personal application.
D19 This is, or will in the foreseeable future be, a *shared workspace application.*
D20 This application is *discrete, describable* and *distinct.*
D21 The *as-used application* is the same as the *as-built application.*
D22 The *service model* will impact TIER(DAL) as follows:

- IF SINGLE-USER THEN WS
- IF CLIENT-SERVER THEN MBS
- IF MULTI-USER THEN MRS

D23 The *application is LAN/WAN dependent* and will use WAN for any inter-tier communication.
D24 The application is *RDBMS-dependent* (invokes D9 also).
D25 The application and workgroup SECURITY requirement is at or below NCSC Level of Trust C-2.
D26 All *four generic sources* are possible sources of specific candidate applications.
D27 Each end-user will use the application only or else the application plus OA. Thus, considering D24 and the SFM, TR = 3000/USER.
D28 Where the default selected tier and processor are already present and available to the workgroup, the *default strategy is to upgrade existing system* and implementation on that target is therefore assumed, providing the upgrade does not drive the processor (and other software on it) up-tier. (Note: this indicator may *not* apply where the workgroup is replicative across many sites unless all of them now have an existing system of this class.)

3.2 Among the indicators set out above, D5 and D18 will in most circumstances dictate the unit tier as the default tier as this is the closest tier to the end-user which still permits full workgroup applications. The next most important indicator is D22, which will usually drive the default choice of processor within the unit tier (i.e. MBS or MRS, depending upon application service model). Only where the service model is single-user will D22 predominate over D5 and D18 in degree of impact on the choice of TIER(DAL).

3.3 Note that the above indicators are *not* intended to serve as a replacement for the main APM nor to involve the analyst in a lengthy consideration of the impacts of re-set indicators (those clearly violated by the application and which must be re-set to another value or assumption). The indicators merely permit the analyst to ensure that the default placement is generally reasonable (at least not ludicrous) given what is already known about the requirement early in the feasibility stage. This does, of course, assume a full and rigorous requirements phase came earlier.

3.4 It will be argued by some that setting DAL at the workgroup tier and thereby forcing one or more indicators to dislodge it to another tier even before the full APM is used is a biased approach. Specifically, it can be suggested that this approach favours the MRS and MBS over WS, midframe and mainframe equipment. However, the following arguments deserve consideration:

- If a displacement model is a valid approach to constructing an APM (and there is no indication that it is not) then it is necessary to displace *from* somewhere – the application must somehow obtain an assumed home base from which we can then consider possible displacement, depending upon the collective impact of all factors considered in the APM.
- Establishing TIER(DAL) with the above indicators permits obvious changes where they are required.
- If the APM is correctly constructed (in both conceptual and logical terms) any bias in the original selection of the value of TIER(DAL) will be netted out anyway.

Step 4: Entering custom factors

4.1 It was clearly recognized, as the APM was being developed, that if the APM was to be used (once partially or wholly automated) by IT to place, for example, 1000 new applications over the next five to seven years that the placement analyses would fall into several categories:

- those that the stock APM (as defined below) could credibly handle
- those in which the stock APM would still function but where its results could be skewed or biased without the introduction (and weighting) of those additional factors of special import or concern to the workgroup, or to IT when considering the workgroup or application, and which are not included in the stock APM
- those in which any 'learning' accomplished by the APM in assessing successive entries of custom factors by different users could add to the quality of the APM decision process

- those in which the APM was seen as clearly inadequate or in which it simply could not function due either to logical (but valid) inconsistencies in the simulands of one or more of the SFM, EMIT, WFM or the APM itself or due to a lack of input data

This sub-section addresses the second and third cases above. In the last case, there is nothing that can be done at this stage in the development to remedy the described situation. The only remedial measures are operational ones during the analysis itself. A case of inadequate information can readily be detected by the analyst and either rectified or the analysis can be declared null and void. However, it will also be necessary for the analyst to have a good understanding of the assumptions about the simuland made by the SFM, EMIT, WFM and the APM itself; otherwise he or she will be unable to detect inconsistencies or 'warpage' in the external reality, which would distort, and hence destroy the validity of, the analysis.

4.2 The following are the steps for entry of custom application placement factors not covered in Step 6 below (and hence also omitted in Appendix B, which will be introduced there).

(1) IT and the client unit or workgroup identify what they *agree* is a valid APM factor, but one which is not included in the stock universe of factors.

(2) The IT analyst would enter important information about the proposed custom factor in a format set out by the model, but which allowed the entry of new entities and the creation of new relationships among new and known entities. Specifically, information entered for a custom factor would include:

- factor identification/name
- factor definition in terms of elements known and unknown and any definable relationships among them
- classification of the factor as being primarily TECHNICAL, OPERATIONAL or ECONOMIC with dual or cross-reference classifications permitted under specified circumstances
- classification of the factor as being ABSOLUTE, RELATIVE or MINIMUM in its tier selection (and hence in its potential tier displacement) impact
- the construction of a prototype rule to permit the analyst to confirm that he or she has constructed an appropriate rule
- upon confirmation by the analyst that the rule is acceptable, calculation of the *range* of potential effects the rule could have – generally this refers to the amount of displacement the rule could

create, and under what, if any, circumstances the rule would be *conclusively absolute*, forcing the APM to divert to an early conclusion and abridging all other rule firing
- entry of special case and commentary information by the analyst

In a circumstance wherein the APM was implemented as an automated tool, there could be an analyst–system dialogue. The system would pose questions to the analyst based on its knowledge of the attributes and behaviour of the entities created or called up (from those already known to it) by the analyst. It would apply general rules and relationships that it knew applied to the known entities and would attempt to resolve inconsistencies or anomalies introduced into its reasoning process by the analyst. It would also highlight any known logical fallacies being introduced by the analyst. Finally, the system would present to the analyst its own generalization hypothesis which it would propose to use in all subsequent analyses to detect circumstances (beyond the current case) where this factor might apply again.

(3) Each custom factor would be classified as being one of the following:

- significant only for this case
- potentially significant for other cases
- probably significant for other cases

(4) When the APM was subsequently run for this specific case the expert system element would attempt to execute the custom factor and would also consider its generalization potential.

Step 5: Information requirements

5.1 The definition of each factor includes the information necessary to its operation. Not all runs of the APM will invoke all factors and the factors actually invoked during a specific case will, of course, determine the actual information intake requirements for that case. For this reason, once any custom factors have been entered, the APM should include a 'dry run' through the custom factors and those stock factors which the analyst reasonably believes will be invoked in the full run. Then, a list of information requirements can be presented by the analyst.

5.2 It is at this stage that the analyst may be enabled to perform the initial cross-check, possibly to determine either that more information must be obtained or else that the APM simply cannot be used in this case due to a critical lack of information. For example, it is unlikely that the full run of the APM could proceed if factors within the ABS classification were lacking data. However, if the data shortfall is related mainly to custom factors just entered, the analyst may wish to re-visit the importance of the custom factors to IT and/or the workgroup.

5.3 All of the information required by the APM must be entered in the prescribed format. Much of this would be interactive and additional information on such items as the characteristics of known system architecture components would be automatically fetched by the APM itself.

Step 6: Run the APM factor matrix

6.1 At this point the analyst is ready to proceed to the full run of the APM. Nothing in the APM, as set out in this book, confines the APM to an automated solution.

6.2 In order to assign a given factor to its respective block in the matrix, it must be classified in two ways, as set out below.

(1) According to the earlier definitions of ABSOLUTE, RELATIVE and MINIMAL. For all stock factors this classification is already made as a part of the APM.

(2) According to the following:

- TECHNICAL, where the factor relates primarily to the design, performance and operation of any technological element included in the system architecture and/or SFM (this broadens the focus to include the rudimentary I/O facilities of the end-user as well as the GUI) – TECHNICAL also refers to the procedural, analytic and other professional aspects of the system development and integration process.
- OPERATIONAL, where the factor relates primarily to the user workgroup characteristics, procedures and practices, their data, WIs and WI flow, as well as application access made by administrator, end-users and any others as well as the interoperation of the application with other applications.
- ECONOMIC, where the factor relates primarily to cost and benefit issues per EMIT, budgetary plans and actual expenditures (past, present or future) or to imputed costs or to the economic opportunity costs of acts or omissions.

6.3 Once a factor's classification under the above two parameters is known, it can be placed in its respective block in the matrix. Figure 5.4 provides the placement of the application placement drivers considered earlier.

	TECHNICAL		OPERATIONAL		ECONOMIC
ABSOLUTE	2.3 3.5 3.8 3.9 4.2 5.1 5.3A 5.3B	5.3C 5.4 5.5 (6.7)	1.3 1.4 2.2 2.4 2.5 2.6J 3.3 3.7	4.7 7.2	1.2 (5.3A) (5.3B) (5.3C) 6.7
RELATIVE	3.1 3.2 (3.8) (3.9) (3.10) 3.12A 3.12B 3.12C 3.12D 3.13 (4.2) 4.3	4.4 5.2A 5.2B 5.2C 5.2D 5.2E 5.2F 5.2G 5.2H 5.2I (5.5) 5.6	1.1 1.5 1.6 1.8 1.9 1.10 2.1A 2.1B 2.1C 2.6A 2.6B 2.6C 2.6D	2.6E 2.6F 2.6G 2.6H 2.6I 2.7 2.8 2.9 2.10 3.4	6.1 6.2 6.3 6.4 6.5 6.6
MINIMAL	3.6 3.11 4.6 7.1		(7.1)		

Figure 5.4 Block assignment of application placement drivers.

This table provides the assignment of application placement drivers (by paragraph number) to the nine-block matrix. Bracketed figures are the secondary assignments of those drivers assigned to two or more blocks; in all but one case this manifestation of a driver is in a lower-value block. Note that sub-sections of application placement drivers have been assigned sequential letter designators not shown in the text.

6.4 Each application placement factor, as derived from a placement driver, can impact one or more of the following attributes of the application:

- tier placement
- service model
- target system within a tier

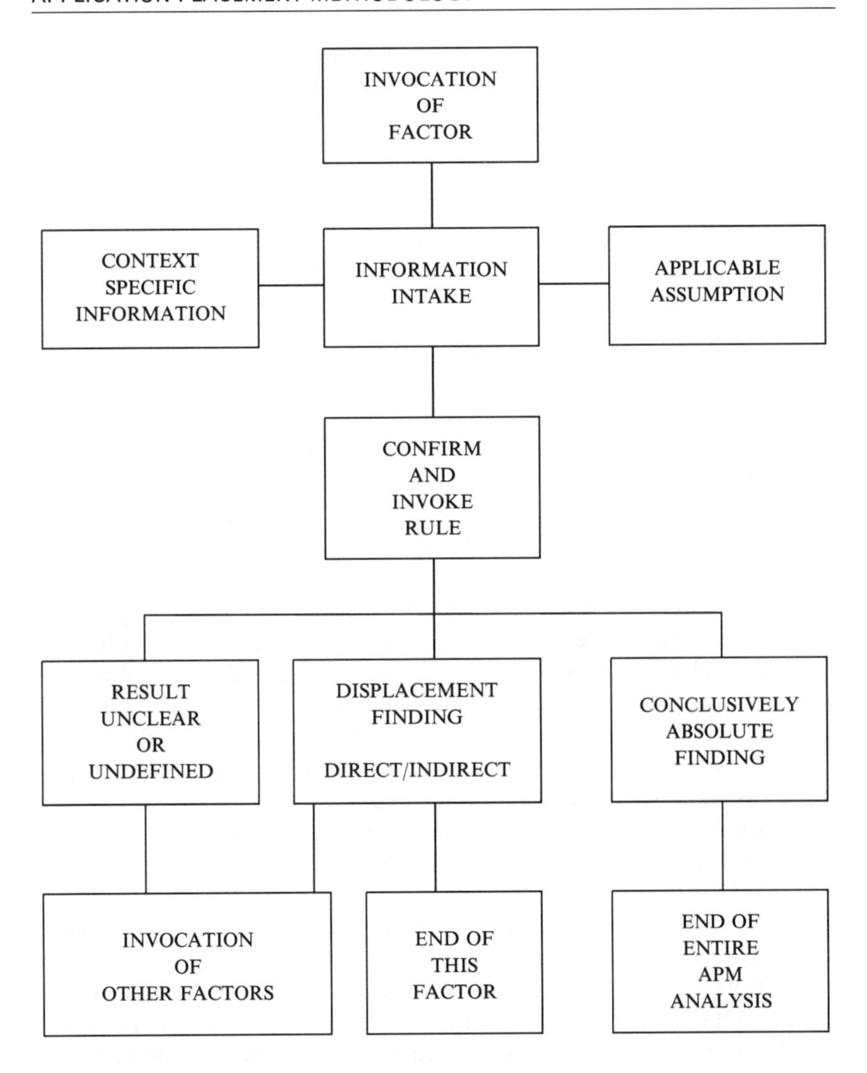

Figure 5.5 Generalized model of APM factor/rule processing.

It will be noted, therefore, that the division of the application placement problem (and hence the mechanics of the decision making) into discrete bundles or portions is somewhat arbitrary.

6.5 Figure 5.5 provides a generalized model of application factor/rule processing within the APM. For each factor, the required information must be obtained and input and it must first be determined whether the factor applies (this is dealt with below).

Where the factor does apply, its constituent rule will be invoked where this is possible and this will result in one of the following outcomes:

- result is unclear or even undefined (in some cases this may invoke another residual or remedial factor)
- result is a displacement finding
- result is a CONCLUSIVELY ABSOLUTE finding which terminates consideration of all other factors

6.6 Appendix B contains the APM factor data specification which will assist in establishing a manual, semi-automated or automated tool for application placement. The first item in the Appendix (following the introduction) is Factor TA-1 (first factor of type TECHNICAL-ABSOLUTE) which addresses the issue of workspace sharing. The rule for this factor will produce either an undefined result or an effect ('E' value) of between 0 and 3. If, for example, a value of 2 was produced in a given case, this would be multiplied by the weighting value for the block in which this factor falls (here, the value is 9); this renders a positive displacement vote of 18 for this particular factor. Similarly, each other factor within this block will have its rule-generated E value multiplied by the block weighting factor of 9. The same procedure is followed for each other block, with the result that each relevant factor for which its respective rule was successfully executed produces a score of a negative integer for down-tier votes, zero for neutral votes and a positive integer for up-tier votes.

If the factor is ABSOLUTE it will either specify or 'de-specify' a given tier. Where a tier is specified, subtraction of the value of TIER(DAL) from this tier will usually render the correct displacement. For example, if a given factor recommends Tier 4 and TIER(DAL) = 2 then the displacement is + 2. This is *implied vectoring*, since the E value must be determined by comparing the selected tier to TIER(DAL). Where a tier is de-specified, no action normally results except in the special case where the de-specified tier is actually TIER(DAL). In such a case it may be desirable to institute either of the following two sets of rules:

- a reserve displacement value to determine whether such rejection of the home tier should result in up-tier or down-tier movement – this value would be specific to the factor in question; or
- a universal reserve displacement value which the analyst could set at the outset of a given run which would override any factor-specific reserve displacement values.

If the factor is RELATIVE or MINIMAL in its displacement strength then DIRECT VECTORING can be employed. The rule will normally not determine a specific choice of tier but will directly determine the tier displacement, often using the function MU> > (MD> >) which causes placement to move up-tier (or down-tier) until a given logical or mathematical condition is fulfilled. The number of tiers moved until the condition is fulfilled is recorded as the E value. For example, the placement called for by a given rule may be first moved up-product within the existing tier, and then as necessary up-tier, until available (i.e. dedicatable) disk storage equals 1.40 times the requirement of the application. Once this condition is fulfilled, the new tier of residency is noted and the difference is recorded as the E value. Of course, up-product movement within a tier leaves E at zero. Note that the MU> > and MD> > functions can be used together (first explore in one vertical direction and then in the other to see if the condition can be fulfilled), sequentially or iteratively.

Thus, the APM algorithm would be as follows:

- move to first (next) block
- move to first (next) factor
- fetch or request required data
- execute rule and determine E value
- multiply E value by block weight and store
- check whether this is last factor in block
- produce 'par table' for factors whose rules were actually executed successfully and for which weighted E values have been stored (see below)
- find the sum of the weighted E values
- compare par score and actual score to determine the recommended placement
- output details of run including factors *not* activated and also rules not able to be executed (which led to factors with undefined results).

You do *not* absolutely require an automated tool to try out and get to know this process. This chapter and Appendix B provide most of what you need to work your way through some sample placements. A spreadsheet approach is another alternative. There are additional benefits from constructing a fully automated tool – one customized to your requirements – but the APM dos not force you to do this. Indeed, you would be foolish to commit resources to automating the APM unless you had already established its worth in solving real-world placement problems within your own organization. Theory is all well and good (and there is, admittedly, much of it in this book), but it is in the practical implementation that it pays off. You will want to ensure that the theory applies at least partially to your situation before committing significant resources to its automation and use.

Step 7: Apply collective displacement vote

7.1 Of course, the result of this process (at the whole matrix level) will almost always be a non-zero integer of some considerable magnitude. A par or default measure is required to permit interpretation of this number. It must also be recalled, however, that with the potential for entry of custom factors unique to each case (where these are truly warranted) no two runs of the APM may produce exactly the same result.

There is a straightforward and context-sensitive means of interpreting the result. Once the universe of *actually invoked* stock and custom factors is known, they form an 'active factor' group. If it were to be assumed that each active factor produced a +1 vote then this can be treated as a 'par'. For each block, given its weighting, a 'block par' can be established and for the entire matrix a 'matrix par' can similarly be established. Table 5.1 presents the result of a universal +1 par for all stock factors only. The matrix par value (if all factors were active and each of their rules voted +1) is thus set at +528, which is also the total of all block pars. For a

Table 5.1 Default (par value) weighting of stock APM factors

BLOCK NUMBER	BLOCK WEIGHT	NUMBER OF FACTORS	FACTOR PAR	BLOCK PAR
1	9	12	+1	+108
2	8	11	+1	+88
3	7	5	+1	+35
4	6	24	+1	+144
5	5	23	+1	+115
6	4	6	+1	+24
7	3	4	+1	+12
8	2	1	+1	+2
9	1	0	+1	0
	Total factors:	86	Stock matrix par:	+528

run of the APM with all of these factors (and only these factors) in use and with all rules executing correctly (i.e. with no undefined results) all negative, zero or positive results can be compared with this par. A final whole matrix result of +998 would point to a nearly two-tier upward displacement; −488 would be interpreted as a one-tier downward displacement and +205 would be insufficient to occasion a displacement. At this stage, normal rounding rules could be employed.

7.2 The above approach is quite straightforward and can be readily diagnosed in the event of an isolated or systematic error. However, the larger question of whether it would be preferable to employ a more sophisticated mathematical method of determining the collective effect of factor votes is beyond the scope of this book. There is also the question of whether statistical or other errors will be within acceptable bounds over some number of runs of the APM. The key issues are set out below.

(1) Should a more sophisticated method of block weighting be employed such that the relative weighting of blocks within the matrix is *not* constant for all runs but is dynamic, being sensitive either to the distribution of factors actually employed (and/or rules successfully executed) or else sensitive to external details related to the specific case?

(2) Should a statistical validity assessment mechanism be introduced into the APM (or else as a separate monitor of the APM's operation) so as to produce one or more measures of confidence regarding the *internal fidelity* of the analysis?

(3) Should some sort of measure of *external fidelity* to the simuland be introduced? (Many advanced simulations include a mechanism to assess the degree of fidelity in simuland representation on a run-by-run basis. The easiest way to do this is to force the analyst to rate his or her confidence in the input data on a datum by datum basis. This would, of course, make preparing and inputting data an extremely tedious activity.)

(The 'simuland' is the real-world thing, process and/or environment which is represented in a model and simulated in a simulation.)

These questions cannot be answered at this time but a few relevant points can be made:

- The APM was developed on an incremental basis and in light of the considerable complexity of the problem it addresses.

- Use of simple ranking of (including one-by-one value incrementation among) points on a continuum or degrees of an attribute is considerably preferable to mere simple generalization about them, but is also often inferior to a more sophisticated approach, where such can be developed.
- Sophistication for its own sake has been carefully avoided – additional complexity almost always introduces additional scope for error (the bush pilot's axiom says that what one does not have cannot break). It was believed preferable to establish a broad conceptual foundation with relatively simple calculation methods, which can always be upgraded later during full-scale development of an automated tool of which the APM is to be a part and without violating the basic principles set out earlier.

Placement verification and validation

There are several possible methods which the analyst can employ, after the APM has been run, to cross-check the recommended placement. These include:

- review of application and user group taxonomy
- verification of APM correct function
- review of key technical, operational and economic issues
- blockage review (does anything else block placement).

The application taxonomy and the user group taxonomy provide important information which can be reconsidered in the event that a placement recommendation seems not to make sense at first glance. These were briefly discussed in the relevant sections above.

Any automated version of the APM should be capable of being set to run in a 'step mode' in which the analyst and the system interactively work their way through one factor at a time and the system presents the information it is using, its assumptions and its calculation logic for the analyst to sanction. While long and laborious, such an interactive run could serve as a backup verification method in the event that a review of the paper output (or electronically stored output file) detailing the vote of each active factor's successfully executed rule was insufficiently revealing.

If output review and a step-by-step re-enactment of the function of the APM both fail to reveal an error when an apparently illogical (or at least intuitively difficult) placement is recommended another option remains. This is a block-by-block review. It is the intent of the APM that the order of precedence on both the vertical and horizontal axes be maintained. Of course, the introduction of a large number of custom factors and/or the non-invocation (or unsuccessful rule execution) of many stock factors could slightly, or radically, alter the respective weighting of the blocks. A review of the case-specific par and as-voted E values in comparison to a stock par chart (i.e. similar to Table 5.1) should provide an indication of the degree to which custom factors and/or factor non-invocation and rule failure have warped or skewed a particular run. This will require a degree of subjective interpretation which will come only with experience in actually using the APM in a variety of circumstances. In other words, APM result interpretation will be partly an art and partly a science. Further, it is doubtful that a more complex specification of the APM will resolve this problem.

6 Planning and implementation model of open computing systems

Overview

Before attempting to unify the three tools developed in Chapters 3, 4 and 5, a brief review may be in order. Those three chapters have provided us with:

- a *system functional model* (SFM), capable of demonstrating the interrelationship of system architecture components as well as alternative application/system combinations capable of fulfilling a given requirement
- an *application placement methodology* (APM), capable of assisting in the placement of applications among the enterprise, area unit and workplace tiers of processing
- an *economic model of information technology* (EMIT) addressing the costs and benefits of open systems

Taken together, these elements can be combined into a comprehensive open systems planning tool, to be called the planning and implementation model for open computing systems (PIMOCS). This tool will enable comprehensive planning for open systems, including the ability to assess the match of candidate applications/systems to the quantitative and qualitative requirements of end-user organizations, the generation of business case documentation and the establishment – over time – of a body of technical, operational and economic precedent.

As shown in Chapter 3, it is possible to construct a system functional model (SFM) using the Open Systems Interconnect (OSI) seven-layer model of communications as a centrepiece. This leads to a graphical representation with three independent stacks, one for each of TECHNOLOGY, COMMUNICATIONS and SUPPORT. Here, 'technology' refers mostly to hardware and software actually used in production systems, 'communications' refers only to communications capabilities, functions and services (as provided to end-user or other clients) and 'support' refers to the provision to end-users and others of all products and services necessary to enable orientation, training and ongoing support of information systems. Within each of the three stacks, each layer is fully dependent upon the capacity, services and facilities of those below it. Each system architecture component can be represented, both in terms of its capabilities and facilities and also in terms of its costs, in the context of the model. Indeed, any given component will map onto some or all layers of one or more of the three stacks. Of course, overlays of individual components are possible to permit 'buildup' of alternative system configurations to support a given application and also alternative application/system combinations to support a given user requirement. Thus, while the model is in the strictest sense a seven-layer model, user requirements and demands are treated as 'Level 8', which must be supported by the seven layers below.

Within a three- or four-tier open system architecture, there are many applications for which the selection of the optimum tier(s) of residency is not a trivial exercise. Unfortunately, most large organizations with legacy systems tended to operate vastly different hardware, system software, packages and languages across different processing tiers. Therefore there exists little literature (or even experience) with respect to how to place (among tiers) an application that could run on some or all available processing tiers. An approach has been developed which considers the organization's operational environment, the end-user community and their current technology base, the nature and extent of the information and the intended application itself, system architectural imperatives and other related factors.

First, a 'default application locus' (DAL) or 'home base' is chosen from among the tiers. Next, individual factors are considered in terms of their degree of importance, on a two-dimensional weighting scale, as follows:

- hierarchy of technical, operational and economic issues
- hierarchy of absolute, relative and minimal impact or certitude in determining a precise tier placement

Finally, the collective individual weighted displacement effects of all applicable factors are applied. Each factor may seek to push the application up-tier or down-tier or confirm the default location; also, factors may seek to alter the service model among single-user, client-server or multi-user. Thus, collectively, the factors will seek either to confirm the default tier placement and service model selection or to alter them. The factors and the logical rules to automate them will require refinement (including further quantification) before the application placement methodology (APM) can be implemented in an automated form.

The APM provides a workable approach within a four-tier architecture, and can be easily generalized to a three-tier architecture. There is no evidence that using a default application locus or 'starting point' in any way biases or prejudices the APM in its function. While this book includes a preliminary identification, definition and quantification of all factors believed relevant to application placement, a more rigorous quantification of some of the factors may be necessary for your particular situation, particularly before attempting to automate the APM.

Based on previous experience, it has been possible to establish an *economic model of information technology* (EMIT). EMIT first considers why local line management would even consider an IT project by allowing them to structure their own arrangement of desired quantitative and qualitative benefits. Additional information is collected about the user workgroup, the application and the information it will manage and the various alternatives. Experience has shown that assessing open systems only in terms of costs and benefits is too narrow and is an inadequate approach. However, open systems do have pure costs and there are also some items which are almost always pure benefits. These can, of course, be offset against one another to the degree that they can be quantified. Open systems have been shown to create many new *opportunities*, which, depending upon the organization's business environment, technology management capabilities and culture, may be exploitable as benefits and/or may impose extra costs.

Although EMIT does not necessarily reinvent or supplant the basic approach and methods used in cost–benefit analysis, it does permit the identification of opportunities arising from open systems, where these opportunities are neither pure costs nor pure benefits. Opportunities can be treated in one or both of two ways:

- They can be assessed in terms of the those negative and positive aspects which *are* quantifiable and then 'netted out' for each year of the project's life cycle and, where desired, reduced to net present value (NPV).
- They can be matched wholly or partially against project investment objectives (whether or not described in quantitative terms) as identified by the intended user organization of the application/system.

The process of developing the *planning and implementation model of open computing systems* (PIMOCS) made clear that there is an even greater degree of interrelationship between the SFM, APM and EMIT 'views' of open system planning than was originally believed to be the case. Full-scale development and full automation of PIMOCS would be a very significant undertaking for any organization, but this would tend to imply a low degree of risk, except where the possible incorporation of an expert system is concerned. Of course, many organizations will find some or all of the manual construction of PIMOCS contained in this book more than sufficient. Also, it is not clear that developing a 'one size fits all' version of PIMOCS would be a good business venture for a software producer, since a significant number of custom factors will be included by most users and automating the incremental logic for these (over and above the stock version) would not be a trivial task. Also, until we have several more years' experience actually using tools like PIMOCS, we will not know the 'experience-validated' benefit or value of using a tool for application placement, versus just taking a wild guess! Intuitively, there is benefit to be had from the approach, but a mathematical proof has not yet been produced; it would depend upon a review of a statistically significant number of case studies of the use of PIMOCS and/or other PIMOCS-like tools. Many SI and consulting firms have developed system planning methodologies, but most of these do *not* take into account the particularities and the special opportunities offered by open systems. They also tend to be designed (often in very subtle ways) to force the user into consuming the consulting services of the originator on an ongoing basis.

If it is desired to produce an automated version of PIMOCS for use within a very large organization (where a series of many application placement decisions that must be made should justify its existence) it would be possible to do so with or without an expert system forming a part of the decision tool. Inclusion of an expert system would permit:

- entry of 'custom' application placement factors, which could be automatically filtered against the existing inventory of stock and (previously entered) custom factors to prevent duplication
- PIMOCS to learn from its experience in actually applying stock and custom APM factors and in generating COB records over time; the ability to apply the best of both the 'common law' and the 'statute law' approaches could be incorporated and utilized

There is every evidence, however, that use of even a paper-only (such as a coil-bound workbook-based) version of PIMOCS – which you can generate yourself from this book – would be highly beneficial in planning for open systems. It would permit you to apply the SFM, APM and EMIT and then to determine the marginal benefits of a basic application over a paper-only version. Certainly, producing a workbook-based version is the quickest and lowest cost way to try out some of the decision tools presented in this book.

Purpose and role of PIMOCS

The general purpose of PIMOCS is to provide an integrated (and automatable) application/system planning tool addressing technology, operational and economic issues related to open systems. The specific purpose of PIMOCS is to proceduralize and unify access to the SFM, APM and EMIT decision tools. The role of PIMOCS is to provide a backbone or structure for the open system planning process, based on a methodological approach which explicitly recognizes the continual interdependence of technical, operational and economic issues throughout the IT project life cycle. Key functions of PIMOCS therefore include:

- collect and store required information
- assess implications of information where possible
- ask relevant questions
- present key opportunities, options and decision points
- advise implications of options and decisions
- record and implement decisions
- generate graphics displays and hard copy output as required

It should be added that PIMOCS neither favours, nor disfavours (nor seeks to make you dependent upon the products or services of) any hardware or software vendor, system integrator or consultant.

PIMOCS should neutralize (or at least minimize) the impacts of bias or prejudice on the part of the project team, the user workgroup and their management and senior management. It should therefore function as a systematic, understandable, predictable and unbiased planning tool which resists attempts (by anyone) to introduce vendor, platform or methodological bias into the planning process. Like the British concept of the 'rule of law' the process itself should be the guiding light, not the views of those who use it.

Nonetheless, good judgement on the part of the participants – exercised within the framework of PIMOCS – is still an important part of the process. Because it is largely procedural, PIMOCS should provide reasonable traceability of where input information came from, how it was used and what the outcome of each such use was, and the collective outcome of all such use. Over time, PIMOCS should permit the establishment of a coherent (and addressable) body of historical information on IT projects, including:

- cost, opportunity and benefit predictions versus actual subsequent experience
- application placements
- introduction and calibration of new system architecture components (generic) or actual vendor products (specific).

Thus, for any application placement or economic analysis there will be, in the future, three alternatives:

- use only the codified rules (the stock format of PIMOCS)
- use the codified rules with extensions or amendments
- establish new processes, relationships and/or rules which will, over time, form a body of precedent offering the potential for a higher form of codification.

Structure and use of PIMOCS

Table 6.1 provides a map of the executive (EXEC), SFM, APM and EMIT 'streams' of PIMOCS matrixed against the traditional system development life cycle (SDLC) phases. Figure 6.1 provides a top-level hierarchical structure for PIMOCS, which has been established according to the same phases. Note that PIMOCS concentrates largely on the phases up to and including the business case, contributes nothing during the definition, design, development and integration/commissioning phases and then is used again at the review stage. Table 6.2 identifies individual functional modules, which are functionally described in Appendix D. Figure 6.2 provides the data storage format used to profile each architectural component.

Table 6.1 PIMOCS stream applicability by project phase

Phase	EXEC	APM	SFM	EMIT	Other
INITIATION	X	X	X	X	X
CLASSIFICATION	X	X	X	X	X
INVESTMENT OBJECTIVES	X			X	X
FEASIBILITY	X	X	X	X	X
BUSINESS CASE	X	X		X	
DEFINITION					
DESIGN					
DEVELOPMENT					
INTEGRATION COMMISSIONING					
REVIEW	X	X	X	X	X

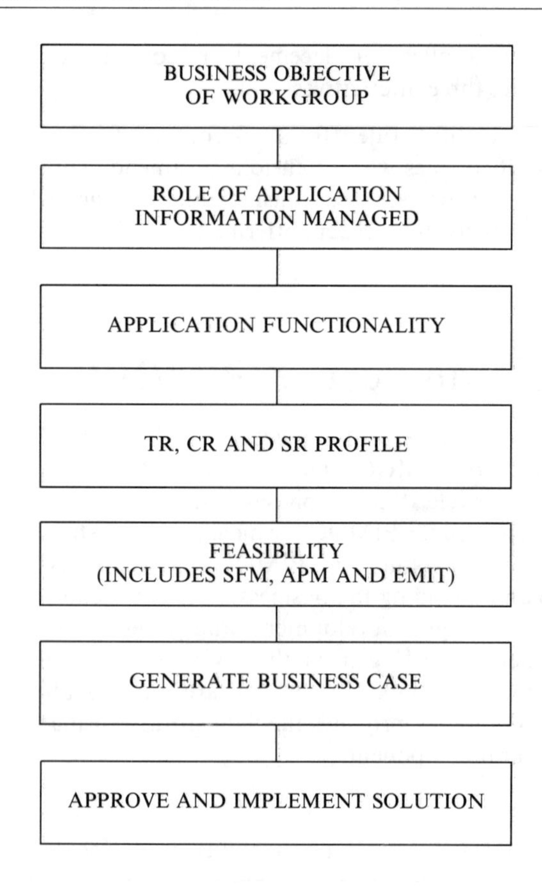

Figure 6.1 Key elements of PIMOCS.

The PROJECT INITIATION (IN) phase involves collection of information about the user workgroup environment and the nominal requirement for the application. Key organizational information is also collected.

The PROJECT CLASSIFICATION (CL) phase addresses the selection of acquisition model, application service model, the general scope of candidate applications and technologies and the analytic approach. All of these classifications should be made as well as the nature (and staging) of various elements of downstream analysis.

The REQUIREMENTS (RQ) phase (that part lying within PIMOCS) provides the intake and filtering of information from either a conventional requirements analysis or a workflow methodology (WFM)

Table 6.2 Phase/stream map of software modules (see Appendix D for descriptions of each module)

Phase	SFM	APM	EMIT	Other
Project Initiation	USRENVT NOMREQ	WGENVT	PROJ ECOENVT	CONSTR CHTR
Project Classification	PROGPLAT COMPONENT SVCEMDL SYSARCH	TIEREXP DEPLSTR USRTAXN	STNDG PLGENVT SYSENVT MOTIVE ANALTYPE	KILL PROSPECT
Requirements	REQTRXL TR CR SR	WORKENVT APPLTAXN	SUPTENVT	GENREQ WFM REQSTMT
Investment Objectives			IMPLASSUM INVCLX OA	CAP-1
Feasibility	APPLCHAR BUILD MATCH	APPLASSUM SEQ SVCEMDL2 DAL CUSTOM INFOREQ PRERUN MATRIX DISPL VERVAL EXTSIT REVIEW	COSTR TWI WICOST TACCOST STRCOST TACOPP STROPP TACBEN STRBEN ESTIM-1	BESTAPPR CAP-2
Business Case	CNFG ACQPROC CNDLIST		ESTIM-2 CAP-3 BUDGET	

Basic Information

GENERIC NAME: ACRONYM:

VENDOR: DESCRIPTION: SUPPORTS:

PEER1: CLASS: PEER2:

REQUIRES:

SUPPLY MODE:

Cost Information

CAPITAL

Corporate Planning/Acquisition/Management .... $
Purchase/Delivery ....
Installation/System Integration/Commissioning ....
Orientation/Training ....

TOTAL CAPITAL .... $

OPERATION AND MAINTENANCE

Corporate Support .... $
Annual Licence Fees ....
Lease/Rental ....
Software Support/Upgrade ....
Hardware Maintenance ....
Administrator Salary and Related O&M ....
Adminstration or Facility Management Contract ....
Supplies and Expendables ....
Re-location/Re-installation ....
Electrical ....
Dedicated facility or Support Equipment ....
Other (   ) ....
Other (   ) ....

TOTAL O&M .... $

Key Attribute/Characteristic Value Summary

ATTRIBUTE	VALUE	TS	CS	SS

Component stack profiles

<table>
<tr><th colspan="3">TECHNOLOGY
(TS)</th><th>COMMUNICATIONS
(CS)</th><th>SUPPORT
(SS)</th></tr>
<tr><td rowspan="3">PACKAGE
( )</td><td>APPLICATION
( )</td><td rowspan="2">APPLICATION
( )</td><td>APPLICATION
( )</td><td>SELF
( )</td></tr>
<tr><td rowspan="2">PACKAGE
( )</td><td>PRESENTATION
( )</td><td>SPECIALIST
( )</td></tr>
<tr><td>LANGUAGE
( )</td><td>SESSION
( )</td><td>LANA/MRSA
( )</td></tr>
<tr><td colspan="3" rowspan="2">OPERATING
SYSTEM
( )</td><td>TRANSPORT
( )</td><td>USER-ASSIST
( )</td></tr>
<tr><td>NETWORK
( )</td><td>VENDOR
( )</td></tr>
<tr><td colspan="3" rowspan="2">HARDWARE
( )</td><td>DATA LINK
( )</td><td>CBI
( )</td></tr>
<tr><td>PHYSICAL
( )</td><td>DOCUMENTATION
( )</td></tr>
</table>

Figure 6.2 Architecture component profile.

(WFM) analysis. This provides further information not only about the nature of the information managed, but also about the likely support environment. At this stage, the requirements must be translated (at least in terms of a good working estimate) into the TR, CR and SR point score requirements utilized by the SFM in building alternative candidate solutions and used by all components of PIMOCS in evaluating them. Thus, a statement of requirements (in terms meaningful to PIMOCS) is produced.

At the INVESTMENT OBJECTIVES (IV) phase a number of assumptions must be made about the actual implementation environment, and preparations are made for a first major project review. Also, if office automation (OA) special mandating – in terms of project justification – is in effect it is invoked to assist in justifying the intended acquisition.

It is at the FEASIBILITY (FS) stage that PIMOCS performs the majority (although by no means all) of its critical functions. Indeed, for many SDLC approaches PIMOCS will perform the majority of the work necessary to complete this phase. PIMOCS is not a replacement for a conventional feasibility study, but it could become a very important component of such a study because:

- It ensures that no platform, tier, operating system, service model or other bias is allowed to creep into the analysis.
- It ensures that all relevant matters are considered.

Alternative candidate solutions (be they applications, systems or more likely application/system combinations) are 'built up' using the SFM and then application placement is performed. It must also be decided whether to economically assess *all* candidates or only the option(s) shown by a conventional feasibility analysis to be able to fulfil all technical and operational requirements. It is usually necessary to construct a full business case if the project is to progress to actual development and implementation.

Data record structures

This section sets out the names and preliminary record structures for the main data sets to be employed in PIMOCS. It has been assumed that where PIMOCS, or a part of it, is automated a full RDBMS will be employed. Of course, there will also be many standalone (temporary or permanent) variables and data sets used in the analysis. A definition for PIMOCS which unduly restricted the design would needlessly reduce the scope for creativity at the later stages.

The following data sets have been identified for use within PIMOCS:

WI	Work item data
COMPONENT	System
CONFIG	Candidate application/system combinations
COB	Cost–opportunity–benefit records
FACTOR	Application placement factors/rules
PLCMT	Application displacement findings
WG	Workgroup environment data
REGISTRY	Software registry application profiles
APPL	Candidate application profiles
ACQ	Acquisition procedures
VENDOR	Vendor/product data
OBJECT	Project objectives

Appendix C provides preliminary record structures to be used as a starting point for constructing PIMOCS data sets to meet your own requirements. You will probably want to extend them to take into account other items important to your organization and/or critically required by your SDLC of choice.

Definition statements

The definitions of each PIMOCS functional module appear in Appendix D. These modules represent the totality of PIMOCS function. The order of presentation of these modules is the order in which a manual implementation of PIMOCS might be set out, as in, for example, a hard copy, coil-bound workbook. The framers of an automated version will have considerably more freedom in determining the sequential (and other) relationships among the various modules.

User interface

For the guidance of anyone seeking to develop an automated version of PIMOCS, the user interface for PIMOCS should be divided into at least five distinct screen areas, assuming a large bit-mapped CRT and a windowing presentation environment. The areas, shown in Figure 6.3, are as follows:

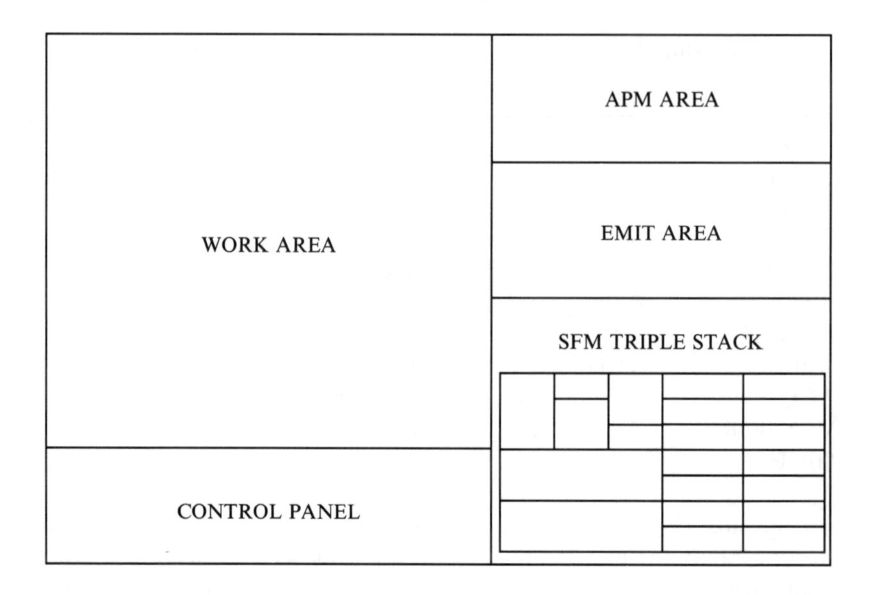

Figure 6.3 PIMOCS conceptual screen layout.

- WORK AREA: for display of options and menus available to the analyst as well as for temporary display of the SFM triple stack.
- CONTROL PANEL: for formulation of commands and for confirmation of selected menu items.
- APM AREA: for display of the APM implications of actions as they are being taken in the CONTROL area.
- EMIT AREA: for display of summaries (or portions) of COB records being impacted as actions are being taken in the CONTROL area – an alternative display is a running 'net result' of a background COB analysis running for the candidate solution now under consideration and being modified
- SFM TRIPLE STACK: for display of the current APPL, COMPONENT or CONFIG item being dealt with, including the following possible colour coding:
  - Green for non-constricted layers within each stack
  - Gold for layers within each stack that are suspected of being constricted by, or may be incompatible with, a version or standard attributed to another item
  - RED for layers known to be incompatible as proposed, or which would not be adequately supported by those below them.

Input/output of model

The following data sets can be treated as default input to PIMOCS:

WI
COMPONENT
FACTOR
WG
REGISTRY
ACQ
VENDOR
OBJECT (default set only)

Within an automated implementation of PIMOCS, working or informal output should be producible at any time by use of standard print commands and/or the PRINT hot-key. Check output of variable and data set field values should be commandable at any time, as should full data set dumps and module transaction histories and results. The nature and detailing of output is, of course, dependent upon the requirements of the organization implementing PIMOCS.

Degree of automation of model for your analysis

This book cannot adequately answer the question of what would be the optimum degree of automation of a tool such as PIMOCS for your organization. As cited above, a totally manual workbook approach, a semi-automated (spreadsheet-based) or a fully automated package (including simulation and AI content) are all possibilities. Key factors which impact this decision can, however, be noted here. They include:

- frequency of use of the tool
- average cost of each application and/or system implementation
- available resources for planning activities
- priority of open systems in the mind of senior management
- availability of IT staff who are both competent analysts and willing simulationists

Conclusion

This chapter is perhaps a bit anti-climactic, since it does not require a lot of detail to tie together the three analytic approaches developed in the previous three chapters. However, Appendix D provides a backbone for the construction of a manual, semi-automated or automated analytic tool which draws on the SFM, APM and EMIT components.

7 Conclusion

Forming realistic expectations of a planning methodology

The material presented in this book is based on my own experience, and on that of some of the many IT professionals with whom I have interacted over the years. That, of course, does not make anything herein either gospel or even a cardinal truth. What this book provides are some concepts and some tools. There are many ways to take tyres off my car; using the right tools will make it easier, make the reverse process possible and leave all else as it should be. Using a blow torch or a grenade will be just as effective in removing tyres (if it is correctly aimed or placed, as the case may be) but may not amuse my neighbours. Selecting not only the right tool, but also the right means of using it, is crucial.

For smaller and mid-sized organizations, the contents of Chapters 3, 4 and 5 should be used as idea-generators and, at best, as guidelines. Chapter 6 may be of less use to a small organization because unification (and formal codification) of the SFM, APM and EMIT is only useful if one or more of the following conditions is true:

- You will launch a sufficient number of individual open systems projects that having an integrated *and consistent* means of looking at the technological, operational and economic aspects of each will save you time, trouble and (hopefully) money.
- You will be acting as an integrator or consultant and you (bravely) believe you can fashion a sufficiently generalized and flexible version of PIMOCS so as to serve the analytic requirements of multiple customers.
- Someone (senior) in your organization has bitterly complained that he or she just cannot understand how IT is making project launch, application placement and system acquisition decisions – often the same person will say that everything should stay on the mainframe until you can *prove* otherwise. When confronting this type of person, PIMOCS should give you enough ammunition and ballistic force to shoot dead an Apatosaurus.

Generally (except possibly for the consultant scenario) all of the above situations connote being a part of a large organization.

Modifying the methodology to meet your needs

The biggest decision you have to take in considering SFM, APM and EMIT (and the unifying framework of PIMOCS) is whether to use them at all. If you can justify using PIMOCS, you can probably justify automating a part of it, at least to the spreadsheet level. There are plenty of good simulation and application-builder tools on the market, but the first thing you will need is a good relational database. The entire methodology is very data set-oriented and this is not an accident. This orientation provides good traceability should anything go wrong; entirely procedural tools generally do not.

Size of organization and mission

There are five levels of use to which you might possibly put the materials contained in this book. (This restates the above commentary somewhat more formally.)

1. Read for interest but do not apply directly to your work
2. Use as an information source or guideline only
3. Implement PIMOCS in a purely manual form, using this book and a series of workbooks as your decision tools
4. Automate aspects of PIMOCS in which you expect to be entering a lot of data, or for which you would otherwise have to perform highly repetitive (and probably very boring) calculations
5. Fully automate PIMOCS with procedural logic, RDBMS, simulation and AI capabilities

Under no circumstances should you jump straight to Option 5 above. Start with an option no higher than 4. A more logical progression is to start at Option 3 and work up from there as you gain experience and become confident that the approach is useful.

Do not be afraid to modify or add to the approach; indeed, the author would be delighted to hear from those who are able to constructively use PIMOCS and also are able to suggest useful additions to the model.

Current system environment

The scale and nature of your current system environment will have an impact on when and how you should consider using a codified open systems planning tool. Although there are many system architecture and system planning tools on the market, most of them have not been optimized to the open systems environment, because their development occurred *before* UNIX-based, office environment-capable, commercial strength workgroup systems became widely used. Sadly too, many of the existing approaches are designed to increase your dependence upon a hardware or software product or upon the consulting or SI organization who sold or gave it to you.

Cost and criticality of system being planned

If you are intending to implement an application or system (or a succession of them) expected to cost millions of dollars, you likely have more to gain from use of PIMOCS than someone who simply contemplates buying a few UNIX workstations for CAD users.

Number of vendors contemplated

Most mid-sized and large organizations will not realize the positive benefits and opportunities of open systems if they do not move to a multi-vendor environment. PIMOCS assists in dealing with more than one vendor in two important ways:

- It allows vendor-independent (generic) consideration to be given to applications, systems and combined application/systems throughout the analysis.
- It could provide documented evidence that fair and unbiased consideration was given to each candidate at each stage in the event that this is required during a subsequent formal procurement review or even a lawsuit.

Some open systems vendors, particularly in the United States, have adopted a rather strong penchant for suing when they do not win mid-sized or large competitive procurements. While PIMOCS has not been court-tested at this writing, the high degree of modularization would permit the defending IT group to explain more readily how decisions were made, even to lawyers and judges whose technical understanding may be (to put it most politely) distinctly limited. Who did what to whom, when, how and on the basis of what reason or motivation are the stuff not only of Sherlock Holmes' attempts to unravel a crime. These 'whodunits' also form a part of the court case when millions of dollars are at stake and the losing vendor is wailing that it was somehow unlawfully discriminated against.

Time/resources available for planning

Obviously, if your organization can afford to spend only a part of one person's time on system planning, full-scale deployment and use of an automated version of PIMOCS is not a practical alternative. Conversely, a very large IT planning unit may already have tools in place covering part of the PIMOCS function and may elect to use only a part of the model. These are the extreme cases; your reality probably falls somewhere between them.

Conclusion: towards a brighter future

As with the arrival on the scene of a new energy source (whether it be steam, the internal combustion engine, fission, fusion or whatever) it has always taken humankind some time to figure out how best to channel and use such energy. The steam engine was used in a stationary mode for quite a while before its ultimate form, the steam locomotive, emerged. The problem we are now facing in the IT world is that two 'energy revolutions' have occurred at the same time:

- Current *advances in computing power* are turning yesterday's prime hardware into museum pieces at a rate which is very hard even to follow, much less to cope with – the power of processors continues to grow very rapidly.
- We now have the *means to make each component of the system architecture, the communications architecture and the (human) support architecture independent of each other component* by defining the interfaces of each component to those around it – these interfaces are called 'standards' and they are the sources from which workable profiles are derived.

Together, these two new 'powers' or 'freedoms' present some very interesting and challenging questions, such as:

- How much power or energy do I require to do the job?
- How do I know when I have too little, enough or too much power?
- How do I match demand and supply for these powers in an intelligent fashion both now and for the foreseeable future?

When we know the answers to these questions, we will be able to obtain *all* of the potentially positive opportunities and pure benefits from the current 'IT revolution'. That (not inconsiderable) challenge, however, promises to occupy us for some years to come.

Appendix A: Workflow methodology

This appendix presents an outline of a methodological tool which was first developed to provide an important bridge to the understanding of (and realistic planning for) the home workplace. It is capable of being implemented in a purely manual (worksheet) form, as a calculative (spreadsheet) tool or as a complete computer simulation program.

The approach is referred to as workflow methodology (WFM) and sufficient information is provided here to permit development, to your own requirements, of a partially or wholly automated comprehensive workplace simulation (CWPS). This approach can used to simulate the flow of all types of work handled by an organization or unit on a person-by-person (or, more correctly, workplace-by-workplace) basis. Together with PIMOCS, this approach can help you to intelligently plan and implement migration to open systems and also (where appropriate to workgroup and worker situations) to the home workplace.

No simulation method or approach can claim to 'predict' with certainty. However, what CWPS does do is to open up a new approach to the implementation of technology within the organization. It cannot prove that there will be more actual use of opportunities for cooperation, more 'co-thinking' or more individual creativity in the organization, but it can show where the greatest scope or *opportunity* for such would actually exist. In fact, the approach is suitable for an organization with no workplace decentralization, one making moderate use of the home workplace or one now in full-scale decentralization. It thus permits consideration of how the organization would function with certain new technologies deployed, irrespective of actual physical deployment.

In choosing a final set of equipment the organization controls the technology to be actually applied. However, once it is installed that technology will in turn impact how the organization operates. CWPS can therefore highlight instances in which system characteristics might be expected to encourage or even necessitate changes in structure, information flows, the way in which work is defined and allocated, job descriptions and procedural guidelines. These types of changes can themselves be simulated by applying the CWPS approach, allowing further detailing of a preferred scenario.

Description

The simulation must accept as input a wide variety of readily collectable/ derivable information from the end-user element of the organization. For each workplace in the organization the following are required:

- name of workplace occupant
- occupant employee category, of which the default categories are:
  - EXEC (executive)
  - PROF (professional)
  - MGMT (management)
  - ADMN (administrative)
  - SUPV (supervisory)
  - OPER (operational)
- computer and communications equipment provided
- key role or function
- reporting position within the organization
- key words on personal interests, abilities, aptitudes
- information/data normally held at workplace
- function rating by function type (see below)
- fully allocated hourly cost
- functional preference ranking

The user will also provide information on the classes of work item which exist within the organization. These could include, for example, handling outside enquiries, handling internal enquiries, tracking whatever it is the organization produces or manages, and writing letters, memoranda and reports, as well as the performance of other routine tasks and special projects. How are these work items presently generated and by whom? How are they allocated or assigned and then followed up? Optionally, particular employee propensities or behaviour can also be considered, such as working only 80% as hard on Friday, or whatever. More information is also required about the interrelationships and propensities of the work items themselves; for example, if work items type A and B interact, this will result in creation of a new work item type C, while the first two types disappear.

An advanced simulation of this type can employ what is called a 'natural language interface', permitting the user to describe the work items, the workers and the workplaces in common English terms and extracting key nouns and verbs. Whether built as a simple microcomputer-based spreadsheet or as a huge program to be run on a mainframe, the primary purpose of the application program is to *simulate the flow of work items through workplaces.* Thus, the flow of the work within the organization is seen as the totality of the flow paths of all work items through whichever sequences of workplaces they visit. This is intuitive, for several reasons.

1. Most work items will at the very least be generated at one workplace and completed at another. There may of course be work performed upon the work item at many workplaces in-between and even those generated and completed at one workplace may impact other work items.
2. When a given workplace is finished with a work item there are only two possible actions: either pass it on to the next workplace that will have to work on it or else 'sit on it'. The latter action is operationally impractical, because if even a few workers sat on all their completed work items the flow of work of the entire organization would eventually become constipated. If accounting does not pay suppliers they will stop supplying and those who receive and process the supplies will soon be impacted.
3. Each work item (or class of routine work items) follows one or a small number of possible paths through the organization.
4. At any point a work item can spawn one or more children or clone itself into two or more successors. For example, an insurance claim may reach a stage where part of it is paid but the other part is submitted to an arbitration or appeals process.

The simulation by a computer of the flow of many work items through a finite and well-characterized group of workplaces produces the following information:

- Work item history, detailing the path taken by a given work item from the origin to the ultimate destination workplace, as well as what human and/or machine functions were performed upon that work item at each stage.
- Workplace history, detailing which work items passed through a given workplace as well as what human and/or machine functions were performed upon them while they were at that workplace.
- Resource history, detailing the total amount of each type of human resource/activity and machine resource/activity dedicated to the performance of work:
  - upon any work item or class of work items
  - at any given workplace
  - for any resource/activity type.

Basically, these three measures are perpendicular to each other in a three-dimensional sense; they look at the same thing (the organization doing its work) from two different perspectives. From the collection of all work item transaction histories (where the work items went, and when and what was done to them as they progressed through the organization from workplace to workplace) it is possible to derive an organizational total workflow. Because the approach can address (if necessary even record) how much time people who are being paid at various rates and who are using various equipment actually devoted to each work item, it is possible to produce a much more thorough work item costing system. If the total amount of employee time available per day or per year within a given organization is held constant, and the amount of work to be processed by that organization is also held constant, then it follows that the application of technologies which speed up and/or improve the quality of the processing of various classes of work items will create new *free time* envelopes (at various times and in various places) for many employees, in which they may choose to:

- remain idle
- perform residual tasks which otherwise are always put off
- process more routine work items
- be creative in some potentially useful way

In real life, as opposed to our antiseptic theoretical world, actual organizations are frequently called upon to handle the current or even a growing workload with *fewer* human resources, often in times of rapid change and high uncertainty. However, the point is that the organization (or more properly the individual who is the person to whom the many little 'envelopes' of technology-created free time will keep presenting themselves) can now choose how best to apply that free time. In most cases it will be a combination drawn from at least the above four kinds of opportunities. This simulation approach cannot predict that an organization will, if a given combination of new technologies is applied, suddenly become more creative at the individual or the corporate level. However, what it can do is predict that many more *opportunities* for such creativity will be available. When you are doing a drudgery-filled and repetitive job all day, or your in-basket is a foot high and everyone is screaming that you are behind, it is hard to be creative.

At the heart of any CWPS implementation is the workflow model, which comprises five distinct stages. Work items may be interactively input by item or class, they may be residually generated by a certain condition set within the simulation or they may be transaction-generated due to spawning/cloning or to scheduling problems. For example, if there are too many work items of type A contending for access to Joe's workplace some may become critically late, spawning work item type X, which is Personnel's work item of hiring Joe's replacement after he has been fired. Alternatively, X might be the hiring of a supplemental worker. It is recognized that in the real world work items will be generated by external stimuli and by internal stimuli, such as routing or bring-forward operatives, the existence of a given set of conditions, the issue of an instruction or even (for a suggestion-processing system) the production of a suggestion by an employee.

A work item's characteristics (definition) may be driven by simulation user input, by some default data set or by the definition of its pre-clone cohort or pre-spawn parent. Work items are simulated to move among workplaces based upon the characteristics and propensities of the work items themselves, those of the workplace and general policies, rules or guidelines existing within the organization (e.g. all supplier invoices paid within 45 days). Work item allocation will occur by either of two methods:

- INSTRUCTION METHOD, wherein someone uses a decision set to make a decision and handles the work item accordingly
- DEFAULT METHOD, wherein the work item is created by an individual who is the only one who can do it or there are timing, loading or other factors which cause the work item to go to one individual automatically

CWPS can also simulate a method of work allocation not yet widely in use today within organizations: the 'expression of interest' or 'bidding' method, wherein the work item's creator person (or even the work item itself) declares its existence and solicits combinations of employees who would line up their workstations (in the logical sense, not the physical one) to process it. The best single or group 'bid' would win the work item; this process is not unlike the mating ritual in the human or animal kingdom. In transportation companies, for example, union contracts call for 'bidding' of certain runs or assignments. The work item, in such a scenario, is 'strutting its stuff' to see who might like to 'do' it. Such an allocation method, within the obvious limits of not permitting workplaces without the required human or machine resources or qualifications to perform the indicated functions, has many advantages. It is a microcosmic free market system operating right inside the organization, matching requirements and their optimum potential fulfilments and providing employees with a much more varied work life.

Let us consider a concrete example. It becomes obvious to Harry, the chief graphic artist in the publications section of an aircraft manufacturer, that next month's load of graphics production for flight manuals for a new and arcane variant of the firm's airplane will overwhelm his department. Even with overtime there are 20 illustrations (each requiring about five hours of combined human attention/computer-aided drawing/design (CADD) station time) which his group cannot complete. Conventional wisdom says he should then draw from the repertoire of things he knows the non-routine work item called 'find a shop and contract the work out' and then apply it to the 20 drawings. But consider the total energy, environmental and economic (E3) cost of this work item, which includes:

- re-check the contracting-out process (he rarely uses it)
- call around for quotes and best completion time
- select best outside shop
- raise a purchase order and any required cost justification documentation (overcome Finance Department's objections too . . .)
- prepare instructions for graphic artist plus samples
- rush purchase order through Contracts Department
- get the material to the outside shop and brief them
- visit the shop at least once to check on progress
- messenger the materials back from the outside shop and short circuit the sloth of the internal mail system
- check the illustrations assiduously
- accept shop's invoice and chase it through the system when the shop calls 70 days later to demand payment

This involves much angst, midnight oil (extra electricity for the three hours of overtime Harry worked, to be precise), two messenger van trips, one return car trip, a car-parking charge, fifteen external phone calls, seven internal phone calls and the price of a large coffee to soothe and encourage a disrupted (and thus disgruntled) contracts officer to hurry up and issue a PO. Alternatively, Harry could have defined the 20 drawings work item electronically and put it onto the internal network.

Jane is an engineer who has been working too many long evenings but now has a break while QA checks over some of her designs; she is also an artist by hobby. Jane quickly bids and wins the job and completes it in half the time of the above process. She also does a better job because she helped design the airplane in the first place. What is more, even if Jane's department *absorbs* the costs (very often the accountants make the chargeback process more painful, expensive and disruptive than just absorbing it) and also accepts the fact that a highly paid engineer is briefly doing the work of the less well-paid graphic artist, the organization is still ahead both in conventional economic terms and in E3 terms. Jane got some much needed variety, and Harry got a better product both faster and with less worry (about the outside shop doing a good job) and less hassle (with Accounting, Contracts, his boss etc.). Everyone wins except the accountants, a whole lot fewer of whom are going to be required once this catches on.

Within the simulation (and this discussion of CWPS treats only the most basic cases and simplifies in some respects) the performance of work by humans and machines at workplaces, upon work items, is achieved by the application of one or more of 14 functions to the work item. The performance of a function requires the electronic physical connection of the person and the work item; in other words the person, the machine and the work item are 'joined' by the fact that the person is using the machine to do things to the work item. In a conventional organization, the individual can interact directly upon the item (without the machine intermediary), but in a 'paperless' organization, and in a decentralized one, the machine would usually be in the loop, although sometimes the 'machine' is only the telephone. A given person may only do one thing to a work item at one time. Naturally, almost any generic description of all of the types of functions is incomplete; the one provided in this basic presentation of the model naturally relates to an organization primarily concerned with the handling of information, as opposed to hamburgers or mice. The 14 functions are:

- thinking (abstract)
- evaluating
- analysing
- comparing, contrasting, sorting or allocating
- searching/retrieving
- reading/reviewing
- filing/despatching/messaging
- typing/entering/inputting
- computing/calculating/logically processing
- telecommunicating
- meeting (physically/conferencing electronically)
- travelling
- composing

The reason for selecting these particular functions was that it is possible to generalize with reasonable accuracy about the relative effectiveness of individuals within various work categories (within an organization) in performing each of them. Employees (by category) can be rated at anything between 0.00 and 1.00 in terms of effectiveness in performing each of these functions. In those organizations where actual measurement is feasible, there are various approaches and disciplines available to develop such ratings, although the reader is cautioned that several of these disciplines decry each other's approaches as nonsense. In any event, the initial run of CWPS for a given organization (in the pre-technology change mode) will allow comparison of simulated and actual (measured or logically imputed) workflows within the organization. This permits the ratings to be varied so as to calibrate the simulation to the organization being simulated (the thing being simulated is technically called the 'simuland'). In other words, we want to get the computer simulation, which implements a model of how work flows through workplaces (and hence through the organization), to behave as much like the simuland as we can, when considering only the technology now deployed. For our example, suppose that everyone had a mainframe computer terminal on their desk, but there were at this stage no PCs or local area networks (LANs).

It has been well noted that the application of new technology can (does not inevitably, but at least 'can') improve worker productivity. Let us assume that enlightened technoids are now summoned who descend upon the organization. They first actually talk to the end-users and then implement the optimum combination of PCs, input/output peripherals (such as printers and optical scanners/readers), communications and software such that people can now work more effectively. Let us also assume that (unlike most organizations) this one actually invests a lot of money up front in orientation, training and initial end-user follow-up support. In short, they make a good job of implementing the new technology. We still have the group of work items which must flow through the same complex of workplaces, but now we have many new powers available at most workplaces. It will be noted that the application of new technology can enhance the ability of the individual to perform all but three (thinking, evaluating and travelling) of the 14 functions. Whereas a new piece of equipment will generally not be applied to a given function (or workplace) unless it is expected that it will bring some improvement, it is reasonable to assume (in general at least) that when an individual is performing a function on a work item where the aid of this equipment is helpful, that the total elapsed time to perform the work item to the *same degree of quality* will decrease. It takes less time to produce a perfect letter, especially if spelling mistakes are first made or the boss changes a word or two, with a PC with word processing software than with a typewriter.

For each function (for example searching/retrieving), a given combination of equipment and software, both at the workplace and distributed throughout the organization, will, where all other things are unchanged, permit the portion of a work item requiring that generic function to be applied to it to be completed in less time than it took previously. In other words, a combination of equipment and software, either at the workplace or elsewhere permits more rapid application of a function to the portion of the work item requiring it. For example, a given work item calls for five hours in the analysis (AN) mode and that work item comes to a person called 'Smith' who is of the professional (P) category and whose default rating for function AN is 1.00. However, Smith has been pegged on the basis of past history (not being 100%

effective at analysing) at 0.90, so the simulation will calculate that it takes her $1/0.90 \times 5.0$ or approximately 5.5 hours to complete this task. If, on the other hand, Smith is given a computer system which is capable of supporting her performance of function AN, then she could use it to reduce this work by the time reduction factor for that function, which we will here assume is 0.50 for AN. The computer lets Smith (or anyone) perform this type of analysis on average twice as quickly as *that person* otherwise would have. The total time for Smith to perform the function with the computer would therefore be 0.5×5.55, or 2.775, which we would round to the nearest 0.25 hours to render 2.75 hours. (I realize this may seem like an elementary mathematics lesson, but there is method in the madness . . . .)

The total performance of work therefore involves four elements:

- function (AN)
- subject (work item name, information item name, data name)
- workplace (Smith, also 'P')
- several manifestations of time
  - arrival time of work item at workplace
  - expected time to perform AN component of work item
  - actual simulated time (given any Smith-unique information)
  - total elapsed clock time that work item was at workplace
  - departure time of work item

Work items generate:

- expenditure of human and machine resources at workplaces
- flows of information/data into, within or out of themselves
- transaction histories (paths through workplaces)

A total of seven work priority levels has been (arbitrarily) identified as follows:

- urgent/interrupt
- external stimulus/instruction-generated
- condition-generated
- bring forward
- project/program/ongoing
- routine/periodic
- idle/residual/reserve/creative

While all members of the organization will have some work items which fall into each of these priority levels or classes, it will be generally true that the priority-level loading will increase with the seniority and degree of extra-organizational interface of an individual. For example, a filing clerk is less likely to be troubled by a continuous stream of urgent tasks than is a director-general. There are, of course, exceptions: it is clear that administrative assistants are constantly interrupted by telephone calls placed often not to them, but to their principals.

If existing work items of all types can be processed more efficiently within the organization once new technology has been implemented, then it follows that CWPS can be utilized to identify the magnitude of the time envelopes that executives, professionals, managers and support staff members may find freed for other activities. Of course, the many thousands of factors impacting these people defy capture in any simulation; all the model will give us is an approximation. The actual uses to which this 'idle/residual/reserve/creative' time is then put by either Smith or the company president will be a function of organizational policy, individual aptitude and interests and a number of other factors. *While the simulation certainly cannot predict what might thus be created or even that creativity will occur, it can identify where in time and space such new scope for creativity is most likely to be created. This assumes only that the input information is valid, the model can be calibrated and something is known about the degree to which the new equipment can assist workers to more rapidly process routine and well-defined work items more rapidly.*

In flowing a work item through a workplace the simulation asks itself seven key questions on behalf of the simulated 'worker'. These are:

- What do I do first/next (i.e. which is the highest priority work item, and if there are two or more at that level, which one has the most key words of interest to me?)?
- Which of the 14 functions is required, or what sequence of functions is required, to complete the work item?
- What information do I need to apply the first (next) function to this work item and what decision set should I use to act on this information?
- Does the work item definition call for possible co-functioning with another specific individual or class of individuals? In other words, do I need others to help me to perform this function on this work item?
- Will I create, modify, revise or destroy any information as a result of the function(s) being performed on this work item?
- If I have finished all functions that I can perform on this work item, where should it go now, and should I let anyone know that I have finished with it?
- If, in going from one function to another while working on this work item, I drop more than five functions in my preference ranking (i.e. I find myself performing any of my nine least preferred functions on this work item), is there someone else in my category and within my section or unit who I might be able to have perform those functions on this work item?

Clearly, this is simple logic which can be easily constructed; organizations can add additional questions for the simulated worker to ask herself as she performs the work item, with scope for any branches or permutations thus indicated. Of much greater interest, this model also allows the construction of a new, although not immediately tractable, concept; this is the concept of the '*smart work item*'. In the same way that the worker can be simulated to ask herself what she should do next, *the work item can be personified and thus simulated to be asking* itself *the following questions*:

- What is my first (next) functional requirement or bundle of functional requirements that I need to have performed on me by a worker?
- Who should perform that function on me (this may be an individual or category set out in the work item's definition or the definition may actually specify 'Joe' as having to do it)?
- Having arrived at that workplace, how long should I remain there before my priority level causes me to prod the worker to 'do me', or else causes me to migrate elsewhere to seek someone else to perform that function or set of functions on me? In draconian situations the work item might abscond to the boss's workplace to complain about lack of attention from Joe.
- When the worker of my current workplace moves to perform the work on me by applying the appropriate human and machine resources for the first/next function, what is his or her function performance rating on this function (1.00 or less), does he or she have the correct information (and if not did he or she get it), does or should he or she co-function with other workers and does his or her use of any particular function either eliminate or raise a new requirement for a subsequent function?

The user of the model may understand why this decision algorithm is necessary to the design of the simulation; however, it may be less obvious why the concept of a smart work item is raised in relation to an organization's future. However, it is my belief that selective use of expert systems/artificial intelligence to enable smart work items could dramatically increase the operational intelligence of installed workplace systems.

CWPS defines 'co-functioning' as the situation where two or more workers, at their workplaces, utilize a common capability or facility or machine function (one supplied to both of them simultaneously by, or at least at, their workstations) to act together upon a given work item or class of work items. Co-functioning generally occurs when conditions A, B and C below are all true plus one of D, E or F is also true:

A The workplaces are separated by one or no employee category tier levels.
B Both workplaces are 'available' in real time and neither is preoccupied with a higher order task.
C Both workplaces have access to the resources needed to co-function.
D The initiator responds to a piece of creative development information created earlier by another worker and directly addresses himself to that worker to see if the two of them might presently work on this work item together.
E The initiator conducts an 'interest search' which is aimed at locating another individual interested in co-functioning on this subject at this time.
F The work item definition itself already calls for parallel functioning, but co-functioning would reduce *total* human and machine resources expended to complete the work item (in other words there are synergies to working together).

The individual workplace activity profile consists of the work items handled (number by class, hours by class and hours by work item unique identification number) as well as the flow of *information* (not bundled with work items) occurring within and to/from the work item and workplace. The latter includes internal information flows among the specific workplace's own storage and processing resources as well as external information flows between this workplace and all other workplaces both inside and outside the organization. The number of unduplicated bi-directional workflow and information flow axes (among workplaces) within the organization is defined by the expression:

$$FX = \frac{X^2 - X}{2}$$

where:

FX = unduplicated inter-workplace axes of flow
X = number of workplaces in the organization but counting only *one* workplace for workers who have two or more

This workflow and information flow/transaction recording capability and its presentation system can be used to support the analysis and presentation of actual data collected in the field as well as the results of simulation runs. Thus it is possible to use the simulation to predict, and then to measure, actual work and information flows. The size and complexity of the organization, the range of tasks performed, and the degree of work item disaggregation will dictate final simulation complexity and hence the required software environment. The work item and unit designation systems described below can also be used as a link between the field assessment data collection/analysis activity and actual CWPS runs.

The work item description will contain the following items of information:

- priority
- system, condition-set, person or entity which generated the work item (parent may be another work item)
- work item type
- number of work items in this immediate class
- subject or designator
- allocating person or method
- definition (reference number of hours of each human and machine resource/function, with any restrictions applying to each, required to complete work item) plus key words and the archiving/recording method

Apart from this descriptive information, each work item will have a unique sequence number assigned to it and a number of 'current' variables which will change as the work item moves from one workplace to another. These latter variables include:

- *Direction*: name or general employee category of the workplace to which work item is to move next
- *Location*: name of occupant of owner of workplace where work item is *now* being processed
- *Time in basket*: time between arrival at this workplace and first processing in workday hours
- *Elapsed time*: total running time at this workplace
- *Function*: function presently being performed on work item
- *Function/time need*: total time which it is calculated that this worker must spend performing each of the functions which he or she is required to perform upon this work item
- *Function/time spent*: total time which has already been spent by this worker performing each of the functions which he or she is required to perform upon this work item.

Finally, a transaction history can be produced for each work item. The transaction history will record:

- workplace name (i.e. worker's name and position category)
- function performed
- time spent on function
- information sought and used

- information revised
- information created
- total time work item

All units of information created and stored within the organization (whether in digital, hard copy, video, image or other form) may be described in terms of the following:

- description of information
- storage type
- storage location
- information type
- information unit number
- revision number

In summary, then, workflow methodology and CWPS provide both a conceptual and a practical tool – in the hands of a competent analyst and a simulation specialist – which can be used to simulate the current manual or semi-automated flow of work through the organization as it exists now. Then, a simulation can be run for a more automated organization, which is *organized (and actually flowing work items among workplaces) in a way which is oblivious to whether the organization exists under one roof, everyone works at home or anything in-between.*

Appendix B: APM factor/rule data records

Introduction

This section contains a preliminary set of data records providing information on each of the stock application placement factors, and their associated rules, as derived from the application placement drivers set out in Chapter 5.

The following information details the meaning of, and provides an example for, each data element in the record.

FACTOR: TA-1 (This factor is the first one of the TECHNICAL and ABSOLUTE type.)
WEIGHT: 9 (The weighting for this matrix block is 9.)
TITLE: WORKSPACE SHARING (The name of the factor.)
INVOCATION: APPLICATION IS NOT SINGLE USER (when factor is used)
INFO REQUIRED: Any special information required.
RULE: IF (TRUE SHARED WG FUNCTION)=TRUE, THEN . . .
- SYSTEM=XWS=XMBS
- SVCMDL=XSU=XCS
- IF TIER(DAL)=1, E=1, ELSE E=0
- IF SYSTEM=MBS, CHANGE TO SYSTEM=MRS

(This rule says that if the logical answer to the question 'Is there true shared workgroup function?' is 'YES' then the system cannot be workstation or MBS. Further, the service model cannot be single-user or client-server. If the tier of Default Application Locus (DAL) is Tier 1 (workplace) then the displacement is set to +1, otherwise to 0. If the original system was MBS it is to be changed to MRS. Of course, migration above Tier 2 would connote a larger system. In some cases the function 'move up until condition satisfied' (MU> >) or 'move down until condition satisfied' (MD> >) will be employed.)

The following abbreviations are used throughout the data records:

SVCEMDL Service model
ACQMDL Acquisition model

SU	Single-user
CS	Client-server
MU	Multi-user
WI	Work item
WP	Workplace
WS	Workstation
MBS	Micro-based system
MRS	Midrange system
MDF	Midframe (a.k.a. ALS = area level system)
MF	Mainframe (a.k.a. ELS = enterprise level system)
APPL	Application
REG	REGISTRY
MOD	MODIFY
TPY	THIRD-PARTY
DEV	DEVELOP
SCE	SOURCE
TGT	Target
TIER(DAL)	Tier of DAL
TIER(TGT)	Tier of target system
ABS	ABSOLUTE
REL	RELATIVE
MIN	MINIMAL OR MINIMUM (ALSO 'MINIMIZE!')
MAX	MAXIMAL OR MAXIMUM (ALSO 'MAXIMIZE!')
XNN	Not NN
!NN	Must be NN
IM	Information management

Tier values are as follows:

ENTERPRISE	4
AREA	3
UNIT	2
WORKPLACE	1

```
FACTOR:  TA-1   WEIGHT:  9 TITLE:  WORKSPACE SHARING
INVOCATION:   APPLICATION IS NOT SINGLE USER
INFO REQUIRED:  SVCEMDL
RULE:  IF (TRUE SHARED WG FUNCTION)=TRUE, THEN . . .
                 - SYSTEM=XWS=XMBS
                 - SVCEMDL=XSU=XCS
                 - IF TIER(DAL)=1, E=1, ELSE E=0
                 - IF SYSTEM=MBS, CHANGE TO SYSTEM=MRS
***************************************************
```

```
FACTOR:  TA-2   WEIGHT:  9  TITLE:  REGISTRY HOME TIER BIAS
INVOCATION:    GENERIC SOURCE = REGISTRY
INFO REQUIRED:  REGISTRY APPLICATION TIER RESIDENCY = TIER(A)
RULE:  E=TIER(A)-TIER(DAL)
**************************************************
FACTOR:  TA-3
INVOCATION:    APPLICATION REQUIRES TPS CAPABILITY
INFO REQUIRED:  TPS REQUIREMENT SPECIFICATION
RULE:  IF REQ=TPS, THEN
                 – SYSTEM=XWS=XMBS
                 – IF TIER(DAL)=1, E=1
                 – IF SYSTEM=MBS, CHANGE TO SYSTEM=MRS
**************************************************
FACTOR:  TA-4
INVOCATION:    APPLICATION REQUIREMENT
INFO REQUIRED:  DEGREE OF PARALLEL PROCESSING
                THROUGHPUT REQUIREMENT
RULE:  IF INVOKED, THEN . . .
                 – SYSTEM=XWS
                 – IF TIER(DAL)=1, E=1
**************************************************
FACTOR:  TA-5   WEIGHT:  9  TITLE:  DIFFICULTY TO PORT
INVOCATION:    GENERIC SOURCE = X DEVELOP
INFO REQUIRED:  DIFFICULTY TO PORT - INTERIM SCALE
                   10   TOTAL RE-WRITE
                    5   CHANGE LANGUAGE VERSION
                    4   SOURCE CODE PORTABILITY
                    3   ENHANCED SOURCE CODE PORTABILITY
                    1   BINARY COMPATIBILITY / HARDWARE MASKING
                    0   FULL BINARY COMPATIBILITY
                TIER-SPECIFIC BARRIERS TO PORT FOR THE THREE TIERS ON
                WHICH THE APPLICATION IS NOW **NOT** RESIDENT EQUALS
                DTP(X) FOR TIER (X) WITH THE VALUES DRAWN
                FROM ABOVE SCALE
RULE:  FIND TIER(N), WHERE TDP(X)=MIN, E=TIER(N)-TIER(DAL)
**************************************************
FACTOR:  TA-6   WEIGHT:  9  TITLE: SERVICE MODEL
RULE:  IF SVCEMDL=SU, SYSTEM=XELS
       IF SVCEMDL=CS, SYSTEM=XWS
       IF SVCEMDL=MU, SYSTEM=XWS=XMBS
       IF SYSTEM AND TIER ARE INCONSISTENT MU>> OR MD>> UNTIL
         INCONSISTENCY IS RESOLVED.
```

```
***********************************************************
FACTOR:  TA-7   WEIGHT:  9  TITLE:  PORTABIILTY: SOURCE-TARGET
INVOCATION:    GENERIC SOURCE = REG, TIER(SCE)= X TIER(TGT)
INFO REQUIRED:  SEE TA-5, TR-3
RULE:  IF DTP(SCE-TGT)>5, E=TIER(SCE)-TIER(DAL), ELSE E=0
***********************************************************
FACTOR:  TA-8   WEIGHT:  9  TITLE: PORTABILITY: TARGET-REGISTRY
INVOCATION:    GENERIC SOURCE = REG, TIER(TGT)=1
INFO REQUIRED:  SEE TA-5, TR-3, TA-7
RULE:  WHERE TIER(TGT)=1, E=1, ELSE E=0
***********************************************************
FACTOR:  TA-9   WEIGHT:  9  TITLE:  PORTABILITY: REGISTRY-NEXT
                                                 TARGET
INVOCATION:    GENERIC SOURCE - REG
INFO REQUIRED:  SEE TA-5, TA-3
RULE:  WHERE TDP (REG-TGT)>5, E=TIER(REG)-TIER(DAL), ELSE E=0
***********************************************************
FACTOR:  TA-10  WEIGHT:  9  TITLE:  PEDIGREE OF APPLICATION
INVOCATION:    APPLICATION IS LANGUAGE OR PACKAGE DEPENDENT.
INFO REQUIRED: DEPENDENCY ON PRODUCT Y WHERE Y DICTATES TIER(X)
RULE:  E=TIER(X)-TIER(DAL)
***********************************************************
FACTOR:  TA-11  WEIGHT:  9  TITLE:  FUTURE OF APPLICATION
INVOCATION:    FUTURE ROLE IS KNOWN TO DICTATE CHANGES TO SCALE OR
               TO OTHER CHARACTERISTICS OF APPLICATION AND THE
               ACQUISITION MODEL IS NOT PLATFORM
INFO REQUIRED:  FOR EACH OTHER FACTOR IMPLICATED . . .
                  X=VALUE OF CURRENT REQUIREMENT
                  Y=UPGRADE MADE NOW TO PLAN FOR FUTURE
                    (IN ALL CASES Y>X)
RULE:             MU>>CAPACITY IS SUFFICIENT, NOTE TIER(P)
                  E=TIER(P)-TIER(DAL)
                  FOR TOTAL OF N FACTORS THUS INVOKED . . .
                  E=SIGMA(E(P1 . . . PN)/N)
***********************************************************
FACTOR:  TA-12  WEIGHT:  9  TITLE:  VALUE OPTIMALITY OF HORIZONTAL
                                    PORTING
INVOCATION:    SVCEMDL=WG
RULE:  SYSTEM=!MRS
       E=2-TIER(DAL)
***********************************************************
```

```
FACTOR:  OA-1   WEIGHT:  8  TITLE:  PROJECT DOABILITY
INVOCATION:    USER
INFO REQUIRED:  TIER OF EXISTING SYSTEM=TIER(X), SVCEMDL
RULE:  IF TIER(X)=1 AND SVCEMDL=SU, THEN TIER(Y)=1,
       ELSE, TIER(Y)=2
       E=TIER(Y)-TIER(DAL)
**************************************************
FACTOR:  OA-2   WEIGHT:  8  TITLE:  WORKGROUP RESIDENCY
INVOCATION:    WORKGROUP IS NOT A UNIT AS PER MRS DEFINITION
INFO REQUIRED:  TIER POSITION OF N WORKPLACES IN WORKGROUP
                ORGANIZATION IN THE **ORGANIZATIONAL** SENSE OF
                TIER, NOT THE SYSTEM SENSE
RULE:  TIER(W)=SIGMA(TIER(WP)FOR WP1..WPN)/N; E=TIER(W)-TIER(DAL)
**************************************************
FACTOR:  OA-3   WEIGHT:  8  TITLE:  MANAGEMENT/USER VIEWS
INVOCATION:    USER
INFO REQUIRED:  USER VIEWS, AS RELATED TO GENERIC SOURCE AND HENCE TO
                TIER OPTIONS
RULE:  SUBJECTIVE INTERPRETATION, TO FIND PREFERRED TIER(P)
       E=TIER(P)-TIER(DAL)
**************************************************
FACTOR:  OA-4   WEIGHT:     TITLE:  WI RELATION TO A GIVEN
                                    INFORMATION MANAGEMENT TIER
INVOCATION:    WHEN WI ANALYSIS IS USED
RULE:  TIER(W)=SIGMA(TIER(WI)FOR WI1..WIN)/N
       E=TIER(W)-TIER(DAL)
COMMENT:  HERE, WE ARE CONSIDERING THE TIER RESIDENCY OF THE WORK
ITEMS EMBODIED WITHIN THE APPLICATION TO TIER(DAL). SIGMA REPRESENTS
THE MATHEMATICAL OPERATION OF ITERATIVELY SUMMING WITHIN A DEFINED
UNIVERSE.
**************************************************
FACTOR:  OA-5   WEIGHT:  8  TITLE:  SHARED FUNCTIONING WITHIN A WI
INVOCATION:    IF TWO OR MORE WIs SIMULTANEOULSY FUNCTION (CO-
               FUNCTION) ON ONE DATA ELEMENT, DB RECORD OR DB FIELD
INFO REQUIRED:  TIER MAP OF WI-TO-DATA ELEMENT FUNCTIONAL VECTORS
RULE:  IF INVOKED THEN, SVCEMDL=XCS, SYSTEM=XWS=XMBS
       MU>> OR MD>> AS REQUIRED TO FULFIL ABOVE
**************************************************
FACTOR:  OA-6   WEIGHT:  8  TITLE:  WI FLOW SEQUENCING
INVOCATION:    ALL CASES
INFO REQUIRED:  WI FLOW SEQUENCING
```

```
RULE:  SUBJECTIVE -     DOES WI FLOW INCLUDE, EXCLUDE OR BIAS FOR
                        OR AGAINST ANY SPECIFIC TIER.
                        FAVOURED TIER . . . . . . TIER(F)
                        DISFAVOURED TIER . . . TIER(D)
                        IF TIER(DAL)=TIER(D), E=-1 PROVIDING SVCEMDL
                        AND SYSTEM REQUIREMENTS ARE NOT VIOLATED
                        IF TIER(DAL)=TIER(D), E=0
                        IF TIER(F)=XTIER(DAL), E=TIER(F)-TIER(DAL)
**************************************************
FACTOR:  OA-7   WEIGHT:  8  TITLE:  GENERIC FUNCTIONS - COMPOSING
INVOCATION:   WI's INCLUDE THE 'COMPOSING' FUNCTION
INFO REQUIRED:  DEFAULT TIER RESIDENCY OF SUBJECT DATA—>TIER(X)
RULE:  E=TIER(X)-TIER(DAL)
**************************************************
FACTOR:  OA-8   WEIGHT:  8  TITLE:  APPLICATION IS FOR WORKGROUP
RULE:  IF INVOKED THEN SYSTEM=XWS, IF TIER(DAL)=1, E=1
**************************************************
FACTOR:  OA-9   WEIGHT:  8  TITLE:  REMOTE ACCESS
INVOCATION:   APPLICATION HAS REQUIREMENT FOR REMOTE ACCESS
RULE:  IF INVOKED, SYSTEM=XWS, IF TIER(DAL)=1, E=1
**************************************************
FACTOR:  OA-10  WEIGHT:  8  TITLE:  GENERIC SOURCE PREFERENCE
INVOCATION:   USER
INFO REQUIRED:  PREFERRED TIER = TIER(X)
RULE:  E=TIER(X)-TIER(DAL)
**************************************************
FACTOR:  OA-11  WEIGHT:  8  TITLE:  DIRECTIVE WITHIN
                                 USER GROUP'S LINE ORGANIZATION
INVOCATION: EXISTENCE OF DIRECTIVE
INFO REQUIRED: TIER CALLED UP BY DIRECTIVE —> TIER(X)
RULE: E=TIER(X)-TIER(DAL)
**************************************************
FACTOR:  EA-1   WEIGHT:  7  TITLE:  BUDGETARY CONSTRAINT
INVOCATION:   USER
INFO REQUIRED:  TIER RESTRICTION DUE TO CONSTRAINT -> TIER(X)
                (NORMALLY TIER(X)=(1,2))
RULE:  E=TIER(X)-TIER(DAL)
**************************************************
FACTOR:  EA-2   WEIGHT:  7  TITLE:  PORTABILITY: SOURCE-TARGET
INVOCATION:   GENERIC SOURCE = REG, TIER(SCE)=XTIER(TGT)
```

```
INFO REQUIRED:  SEE TA-5, TR-3
RULE:  INVOKE EMIT TO COMPARE COST TO PORT. WHERE THIS EXCEEDS
       PERCEIVED BENEFIT, APPLICATION IS DEEMED TO HAVE AN ABSOLUTE
       PREFERENCE FOR ITS CURRENT TIER, HENCE:
          E=TIER(SCE)-TIER(DAL), ELSE E=0
*************************************************
FACTOR:  EA-3   WEIGHT:  7  TITLE:  PORTABILITY: TARGET-REGISTRY
INVOCATION:    GENERIC SOURCE = REG, TIER(TGT)=1
INFO REQUIRED:  SEE TA-5, TR-3, TA-7
RULE:  WHERE ESTIMATED COST TO PORT EXCEEDS PERCEIVED BENEFIT,
       APPLICATION IS DEEMED TO HAVE AN ABSOLUTE PREFERENCE FOR ITS
       THEN-CURRENT (IE: TARGET) TIER. HENCE:
          E=TIER(TGT)-TIER(DAL), ELSE E=0
*************************************************
FACTOR:  EA-4   WEIGHT:  7  TITLE:  PORTABILITY: REGISTRY-NEXT
                                                 TARGET
INVOCATION:    GENERIC SOURCE = REG
INFO REQUIRED:  SEE TA-5, TR-3
RULE:  WHERE ESTIMATED COST TO PORT EXCEEDS PERCEIVED BENEFIT,
       APPLICATION IS DEEMED TO HAVE AN ABSOLUTE PREFERENCE FOR THE
       THEN-CURRENT REGISTRY TIER, HENCE:
          E=TIER(REG)-TIER(DAL), ELSE E=0
*************************************************
FACTOR:  EA-5   WEIGHT:  7  TITLE:  VALUE OPTIMALITY OF HORIZONTAL
                                    PORTING
INVOCATION:    APPLICATION IS FOR WORKGROUP
INFO REQUIRED:  N = NPV OF FUTURE PORTING COST (OTHER TIER TO MRS)
                A = COST OF NON TIER 2 APPLICATION
                B = COST OF TIER 2 APPLICATION
RULE:  WHERE A>B, C=A-B
       IF N>C, THEN TIER(X)=2
          E=TIER(X)-TIER(DAL)
*************************************************
FACTOR:  TR-1   WEIGHT:  6  TITLE:  CAPACITY SURPLUS
INVOCATION:    ALL CASES
INFO REQUIRED:  TS(A) = SFM TS SCORE OF CONSIDERED TIER WHICH IS
                DEEMED TO BE TIER(A)
                TR(B) = SFM TR SCORE REQUIREMENT OF APPLICATION
RULE:  MU>>TS(A)=(1.40*TR(B))
*************************************************
FACTOR:  TR-2   WEIGHT:  6  TITLE:  PROXIMITY TO END-USER
```

INVOCATION: ALL CASES

INFO REQUIRED: USER GROUP TIER RESIDENCY ->TIER(WG), SVCEMDL

RULE: IF APPLICATION IS FOR WORKGROUP, MD>>TIER=TIER(WG), BUT
DO NOT SET TIER=1, ELSE MD>>TIER=TIER(WG)

****

FACTOR: TR-3 WEIGHT: 6 TITLE: TRANSACTION PROCESSING

INVOCATION: APPLICATION REQUIREMENT

INFO REQUIRED: TPS PERFORMANCE REQUIREMENT (TPS) = Z

RULE: MU>>Z(APPL)=Z(TIER)

****

FACTOR: TR-4 WEIGHT: 6 TITLE: PARALLEL PROCESSING

INVOCATION: APPLICATION REQUIREMENT

INFO REQUIRED: DEGREE OF PARALLELISM, THROUGHPUT REQUIREMENT
ESTABLISH SPECIFIC PARAMETER OF PERFORMANCE FOR THIS
APPLICATION AND SET Z TO A VALUE OF THAT PARAMETER

RULE: MU>>Z(APPL)<Z(TIER)

****

FACTOR: TR-5 WEIGHT: 6 TITLE: DISTRIBUTED PROCESSING

INVOCATION: ALL CASES

INFO REQUIRED: SVCEMDL, LOGIC SPLIT AMONG CENTRAL AND DISTRIBUTED
PROCESSING ELEMENTS
TR(B) = REQUIRED CAPACITY FOR EACH
DISTRIBUTED APPLICATION ELEMENT
TS(A) = CAPACITY OF SYSTEM/TIER UNDER
CONSIDERATION

RULE: MD>>TR(A)=(1.40*TS(B)), BUT IF SVCEMDL=XSU, SET TIER>1

****

FACTOR: TR-6 WEIGHT: 6 TITLE: TECHNOLOGY - CPU REQUIREMENT

INVOCATION: ALL CASES

INFO REQUIRED: SELECT CPU POWER MEASURE DESIRED
A = APPLICATION CPU POWER REQUIREMENT
B = CPU POWER RATING OF TIER UNDER CONSIDERATION

RULE: MU>>B=(1.40*A)

****

FACTOR: TR-7 WEIGHT: 6 TITLE: TECHNOLOGY - MEMORY REQUIRED

INVOCATION: ALL CASES

INFO REQUIRED: A = APPLICATION MEMORY REQUIREMENT (MB)
B = SYSTEM DEDICATABLE MEMORY (MB)

RULE: MU>>B=(1.40*A)

****

FACTOR: TR-8 WEIGHT: 6 TITLE: TECHNOLOGY - STORAGE REQUIRED

```
INVOCATION:     CALL CASES
INFO REQUIRED:  A = APPLICATION STORAGE REQUIREMENT (MB)
                B = SYSTEM DEDICATABLE STORAGE (MB)
RULE:  MU>>B=(1.40*A)
**************************************************
FACTOR:  TR-9   WEIGHT:  6  TITLE:  TECHNOLOGY - RDBMS SUPPORT
INVOCATION:     APPLICATION REQUIRES RDBMS
INFO REQUIRED:  DFXTS = DATABASE FEATURE RICHNESS (1-5)
                DBCAP = DATABASE THROUGHPUT CAPABILITY (SEE BELOW)
                DBPERF = DFXTS*DBCAP
RULE:  MU>>DBPERF(TIER)=(1.20*DBCAP(APPL))
COMMENT:  SELECT UNIT OF MEASURE FOR DB WORK PROCESSING CAPABILITY OR
THROUGHPUT CAPABILITY APPROPRIATE TO SITUATION. NORMALIZE SCALE OF
DBCAP TO 1-100. DBPERF ADDS A FURTHER DIMENSION TO THE SFM-DICTATED
TS=3000 RATING FOR SFM TIERS 5-6. THE MEASURE OF RDBMS FEATURE-
RICHNESS IS OF COURSE SUBJECTIVE, SO USE YOUR OWN ORGANIZATION'S
EXPERIENCE TO MAKE THIS DETERMINATION. HINT: ORACLE AND SOME OF THE
MORE MODERN NATIVE MAINFRAME DATABASES SHOULD BE RATED AS A '5'.
**************************************************
FACTOR:  TR-10  WEIGHT:  6  TITLE:  CLIENT-SERVER OPTIMIZATION
INVOCATION:     SVCEMDL=CS
INFO REQUIRED:  TIER(C) = CLIENT'S TIER OF LOCUS
                TIER(S) = SERVER'S TIER OF LOCUS
RULE:  AS REQUIRED,     MU>>TIER(S)=TIER(C), OR
                        MD>>TIER(S)=TIER(C), BUT
                        DO NOT SET TIER(S)=1
**************************************************
FACTOR:  TR-11  WEIGHT:  6  TITLE:  DIFFICULTY TO PORT
INVOCATION:     GENERIC SOURCE = XDEV
INFO REQUIRED:  SEE TA-5
                Z = DESIRED LEVEL OF DIFFICULTY TO PORT SCE->TGT
                (DERIVE Z FROM TA-5 SCALE)
RULE:  MU>>DTP(TIER)=Z OR DTP(TIER)<Z
**************************************************
FACTOR:  TR-12  WEIGHT:  6  TITLE:  ACQUISITION CLASSIFICATION
INVOCATION:     ALL CASES
INFO REQUIRED:  ACQMDL
RULE:  IF ACQMDL=PLATFORM AND THE ACQMDL SELECTION IS MADE EARLY AND
       IS NOT ELS OR ALS, AND IF THIS IS A WG APPLICATION,
       IF SVCEMDL=XSU, MD>>TIER=2
       IF SVCEMDL=SU, MD>>TIER=1
```

```
*****************************************************
FACTOR:  TR-13  WEIGHT:  6  TITLE:  TYPE OF REQUIREMENT ANALYSIS
                                    PERFORMED EARLIER
INVOCATION:   ALL CASES
INFO REQUIRED:  TYPE OF ANALYSIS (WFM OR NOT)
RULE:  IF CONVENTIONAL ANALYSIS USED, OFFSET WITH E=-1
       IF WFM ANALYSIS USED, OFFSET WITH E=1
*****************************************************
FACTOR:  TR-14  WEIGHT:  6  TITLE:  APPLICATION LINES OF CODE (LOC)
INVOCATION:   ALL CASES
INFO REQUIRED:  A =   LOC
                B =   PRACTICAL LOC MAXIMUM OF ALLOCATED PORTION OF
                      TARGET SYSTEM ON TIER CURRENTLY UNDER
                      CONSIDERATION
RULE:  MU>>B=(1.25*A)
*****************************************************
FACTOR:  TR-15  WEIGHT:  6  TITLE:  APPLICATION LOGIC COMPLEXITY
INVOCATION:   ALL CASES
INFO REQUIRED:  REQUIRES A SCALE OF APPLICATION LOGIC COMPLEXITY
                EXTENDING FROM 1-10 IN TERMS OF ACTUAL SYSTEM
                RESOURCE CONSUMPTION. SCALE SHOULD BEAR LINEAR OR
                LOGARITHMIC RELATIONSHIP TO SFM TR SCORE. EACH SYSTEM
                MUST BE RATED IN ITS ABILITY TO HANDLE THIS DEGREE OF
                COMPLEXITY IN TERMS MORE PRECISELY HONED TS MEASURE
                USED ONLY FOR THIS FACTOR.
RULE:  QUALITATIVELY SELECT PLATFORM BASED ON LOGIC COMPLEXITY
COMMENT:  EITHER A MACHINE CYCLE, OPERATING SYSTEM PROCESS OR A
          WORKLOAD-BASED MEASURE (PER VARIOUS BENCHMARKING
          APPROACHES) CAN BE USED HERE, PROVIDING IT IS USED
          CONSISTENTLY IN ASSESSING THE REAL APPLICATION
          REQUIREMENTS AND SYSTEM PERFORMANCE CAPABILITIES. NOTE
          THAT USING TS WILL LIKELY RENDER A TS SCORE SLIGHTLY IN
          EXCESS OF THE OTHERWISE NORMALLY CALCULATED FOR A GIVEN
          SYSTEM AS PER THE SFM. THUS, THE TS SCORES DERIVED FOR THIS
          FACTOR CAN ONLY BE USED FOR THIS FACTOR AND ARE NOT PORTABLE
          AMONG FACTORS.
*****************************************************
FACTOR:  TR-16  WEIGHT:  6  TITLE:  APPLICATION EXTERNAL CALLS
INVOCATION:   ALL CASES
INFO REQUIRED:  A = NUMBER OF EXTERNAL CALLS - SAME TIER
                B = NUMBER OF EXTERNAL CALLS - HIGHER TIER
                C = NUMBER OF EXTERNAL CALLS - LOWER TIER
```

T = A+B+C

RULE: IF(B2)>(A2+C2), E=1
IF(C2)>(A2+B2), E=-1
ELSE E=0

****

FACTOR: TR-17 WEIGHT: 6 TITLE: APPLICATION INTER-OPERATION WITH OTHER TIERS

INVOCATION: ALL CASES

INFO REQUIRED: WI(N) = WI CLASS WITH EXTERNAL UP-TIER RELATIONSHIP OR DOWN-TIER RELATIONSHIP
CLASS(N) = NUMBER OF WI's IN CLASS N
A = SIGMA (WI(N)*CLASS(N)) FOR UP-TIER
B = SIGMA (WI(N)*CLASS(N)) FOR DOWN-TIER
C = SIGMA (WI(N)*CLASS(N)) FOR SAME-TIER
T = A+B+C

RULE: IF(B^2)>(A^2+C^2), E=1
IF(C^2)>(A^2+B^2), E=1
ELSE E=0

****

FACTOR: TR-18 WEIGHT: 6 TITLE: APPLICATION DATA VOLUME

INVOCATION: ALL CASES

INFO REQUIRED: APPLICATION SPECIFICATION
OPERATIONAL PLAN INFORMATION
A = APPLICATION NON-VECTORED DISK I/O (MB/HR)
B = TARGET SYSTEM RATED DISK I/O (MB/HR)
C = PORTION OF SYSTEM I/O CAPABILITY DEDICATABLE TO APPLICATION ON A CONTINUOUS BASIS

RULE: MU>>B=(1.5*A)

****

FACTOR: TR-19 WEIGHT: 6 TITLE: APPLICATION VECTORIZED EXTRA-TIER I/O

INVOCATION: ALL CASES

INFO REQUIRED: A = VECTORED I/O (MB/HR) - SAME TIER
B = VECTORED I/O (MB/HR) - HIGHER TIER
C = VECTORED I/O (MB/HR) - LOWER TIER
T = A+B+C

RULE: IF(B^2)>(A^2+C^2), E=1
IF(C^2)>(A^2+B^2), E=1
ELSE E=0

****

FACTOR: TR-20 WEIGHT: 6 TITLE: MACHINE I/O

INVOCATION: ALL CASES
INFO REQUIRED: APPLICATION SPECIFICATION
SYSTEM BENCHMARK DATA
A = EST. APPLICATION MAX AGGREGATE I/O (MB/HR)
B = SYSTEM RATED MAX I/O (MB/HR)
C = PORTION OF SYSTEM RATED I/O DEDICATED (0-1.00)
RULE: MU>>(C*B)=(1.40*A)

****

FACTOR: TR-21 WEIGHT: 6 TITLE: SYSTEM PROCESSES
INVOCATION: ALL CASES
INFO REQUIRED: BENCHMARK DATA
A = EST. MAX APPL CONCURRENT SYSTEM PROCESSES (#)
B = SYSTEM RATED MAX CONCURRENT PROCESSES (#)
C = PORTION OF SYSTEM DEDICATABLE TO APPL (0-1.00)
RULE: MU>>(C*B)=(1.40*A)

****

FACTOR: TR-22 WEIGHT: 6 TITLE: APPLICATION MEMORY REQUIREMENT
INVOCATION: ALL CASES
INFO REQUIRED: APPLICATION SPECIFICATION
SYSTEM BENCHMARK DATA
A = EST. MAX APPL/DATA/JCL ON-LINE MEMORY (MB)
B = SYSTEM RATED AVAILABLE MEMORY
C = PORTION OF SYSTEM DEDICATED TO APPL (0-1.00)
RULE: MU>>(C*B)=(1.40*A)

****

FACTOR: TR-23 WEIGHT: 6 TITLE: APPLICATION FUTURE
INVOCATION: IMPLICATIONS OF FUTURE APPL ROLE (CHANGE OF SCALE)
INFO REQUIRED: SEE TA-11
RULE: USE TA-11 RULE TO DETERMINE TIER(A) WITH CONSIDERATION
OF **FUTURE** REQUIREMENTS.
THE GENERAL CASE IS MU>> UNTIL REQUIREMENTS SATISFIED

****

FACTOR: TR-24 WEIGHT: 6 TITLE: APPLICATION SUPPORT
REQUIREMENTS
INVOCATION: ALL CASES
INFO REQUIRED: AN APPROACH IS REQUIRED FOR THE CLASSIFICATION OF
LEVEL OF SUPPORT. THIS MAY INVOLVE ENHANCING THE SFM
TO TAKE ACCOUNT OF APPLICATION-SPECIFIC NEEDS. THIS
COULD BE USED TO WEIGHT THE SFM SR POINT SCORE
(SR=1000, SR=2000 OR SR=3000) WITH AN INDICATION OF
THE 'INTENSITY' OF SUPPORT NEEDED, PERHAPS ON AN HRS/

```
                YR BASIS PER INSTALLATION OF THE APPLICATION AND/OR
                PER WORKGROUP SITE. THIS COULD BE MULTIPLIED BY THE SR
                RATING TO OBTAIN AN 'SR ANNUAL' OR SRA RATING FOR
                COMPARISON AGAINST A DEFAULT. SIMILARLY, SSA RATINGS
                CAN BE PRODUCED FOR EACH TIER.
RULE:  MU>>SSA(TIER)=SRA(APPL)
************************************************
FACTOR:  OR-1  WEIGHT:  5  TITLE:  WORKGROUP EXT RELATIONSHIPS
INVOCATION: APPLICATION REQUIREMENT
INFO REQUIRED:  TIER(EXT) = TIER OF EXTERNAL APPLICATION
RULE:  IF TIER(EXT)=TIER(DAL), E=0
       ELSE, MU>>TIER=TIER(EXT) OR MD>>TIER=TIER(EXT)
       E=TIER(EXT)-TIER(DAL)
************************************************
FACTOR:  OR-2  WEIGHT:  5  TITLE:  INVESTMENT OBJECTIVE - BRING
PROCESSING CLOSER TO USERS
INVOCATION:   USER
INFO REQUIRED:  USER GROUP MEMBER ORGANIZATIONAL TIER RESIDENCIES
                TIER(X) = SIGMA(TIER(Y), Y=1..N OF MEMBERS)
RULE:  E=TIER(X)-TIER(DAL)
************************************************
FACTOR:  OR-3  WEIGHT:  5  TITLE:  EXISTING EQUIPMENT
INVOCATION:   USER
INFO REQUIRED:  SVCEMDL, ACQMDL OF EXISTING SYSTEM
                TIER(X) = TIER OF EXISTING SYSTEM
RULE:  IF SVCEMDL(EXISTING)=SVCEMDL(DAL), E=TIER(X)-TIER(DAL)
************************************************
FACTOR:  OR-4  WEIGHT:  5  TITLE:  UNIT WORKSITE LOCATION
INVOCATION:   USER OR ANALYST
INFO REQUIRED:  WORKSITE ACCESSABILITY RATING
RULE:  ASSUME THAT SUPPLY OF REQUIRED SUPPORT (SS=A) IS
       'DEFLATED' BY A DISTANCE/ACCESSABILITY FACTOR (DAF) USE
       COMPONENT RATING FOR EACH OF LEVELS 1-7 OF SUPPORT ELEMENT OF
       SFM AND NOTE RATING OF MOST CONSTRICTING LEVEL OF SUPPORT FOR
       INTENDED WORKGROUP LOCATION. WHERE DAF=0.XX, THEN BY
       DEFINITION THE ACTUAL LEVEL OF SUPPORT (HERE, CALLED 'B') IS
       LESS THAN A
          THEREFORE; MU>>SS(B)>SR(APPL)
************************************************
FACTOR:  OR-5  WEIGHT:  5  TITLE:  WG SITE REPLICATION
INVOCATION:   APPLICATION = REPLICATIVE
INFO REQUIRED:  A = EMPLOYEES PER SITE
```

```
            B = SYSTEM USERS PER SITE
            C = APPLICATION USERS PER SITE
            D = CURRENT MANUAL PARTICIPANTS PER SITE
            X = PURE OA POTENTIAL (A-C)
            Y = TOTAL AUTOMATION POTENTIAL (A-B)
            Z = (X+Y)/2 AND N = (C+Z)/2
RULE:  WHERE SVCEMDL=MU, FOR N=1 . . 4, SYSTEM=MBS
                         FOR N=5 . . 50, SYSTEM=MRS
                         FOR N=51+, SYSTEM=ALS
       WHERE SVCEMDL=CS, FOR N=1 . . 30, SYSTEM=MBS
                         FOR N=31..70, SYSTEM=MRS
                         FOR N=71+, SYSTEM=ALS
****************************************************
FACTOR:  OR-6  WEIGHT:  5  TITLE:  WORKGROUP NOT CO-LOCATED
INVOCATION:   WORKGROUP IS NOT CO-LOCATED, NO DISTRIBUTED
              PROCESSING IS CONTEMPLATED
INFO REQUIRED:  WORKGROUP LOCATIONS
RULE:  IF TIER(DAL)=1, E=2
       IF TIER(DAL)=2, E=1
       ELSE, E=0
****************************************************
FACTOR:  OR-7  WEIGHT:  5  TITLE:  SYSTEM GENERAL AVAILABILITY
                                   AND RELIABILITY
INVOCATION:   ALL CASES
INFO REQUIRED:  WI(A), WI(T)         A = WI(A)/WI(T)
RULE: IF A>.30 AND TIER(DAL)=1, E=1
       IF A>.30 AND TIER(DAL)=2, E=0
       IF A>.50 AND TIER(DAL)=2, E=1
       ELSE, E=0
****************************************************
FACTOR:  OR-8  WEIGHT:  5  TITLE:  HARDWARE AVAILABILITY,
                                   RELIABILITY AND REDUNDANCY
INVOCATION:   ALL CASES
INFO REQUIRED:  WI(A), WI(T) APPLICATION REQUIRED RATINGS FOR X, Y, Z
                BELOW SYSTEM RATINGS FOR X, Y, Z BELOW
                X = AVAILABILITY (%)
                Y = RELIABILITY (HRS CONTINUOUS OPERATION MTBF)
                Z = 1 FOR BACKUP, 2 FOR STANDBY, 3 FOR HOT STANDBY
RULE:  MU>>X(TIER)=X(APPL)
       MU>>Y(TIER)>Y(APPL)
       MU>>Z(TIER)=Z(APPL)
```

```
      SELECT HIGHEST TIER FROM ABOVE THREE RESULTS ->TIER(A)
      E=TIER(A)-TIER(DAL)
*************************************************************
FACTOR:  OR-9   WEIGHT:  5  TITLE:  COMMUNICATIONS CONSISTENCY /
                                    RELIABILITY
INVOCATION:    ALL CASES
INFO REQUIRED:  WI(A), WI(T) APPLICATION REQUIRED RATINGS FOR X, Y
                BELOW SYSTEM RATINGS FOR X, Y BELOW
                X = COMMUNICATIONS SUB-SYSTEM AVAILABILITY (%)
                Y = RELIABILITY (HRS CONTINUOUS OPERATION MTBF)
RULE:  MU>>X(TIER)=X(APPL)
       MU>>Y(TIER)>Y(APPL)
       SELECT HIGHEST TIER FROM ABOVE THREE RESULTS ->TIER(A)
       E=TIER(A)-TIER(DAL)
*************************************************************
FACTOR:  OR-10  WEIGHT:  5  TITLE:  WI FUNCTION - ANALYSIS
INVOCATION:    IF WFM USED
INFO REQUIRED:  K = OCCURRENCES OF THIS FUNCTION IN WI CLASS N
                L = NUMBER OF WIs IN CLASS N
                T = NUMBER OF WI CLASSES IN APPLICATION
                FP = SIGMA(K(N)*L(N), FOR N=1..T)
                M = OCCURRENCES OF ALL FUNCTIONS IN WI CLASS N
                TP = SIGMA (M(N)*L(N), FOR N=1..T)
                A = FP/TP
                TR(B) = A*TR(APPL)
                C = ALLOCATION PORTION OF TARGET SYSTEM (0-1.00)
                Z = REQUIRED TS RATING TO PERFORM WORKLOAD FP
                    FROM NORMALIZED BENCHMARK DATA, BASED ON
                    NUMBER AND INTENSITY OF PROCESSES LAUNCHED
RULE:  MU>>(C*TR(B))=((1.40*TS(TGT)*A)+Z)/2
*************************************************************
FACTOR:  OR-11  WEIGHT:  5  TITLE:  WI FUNCTION COMPARING/
                                    CONTRASTING/SORTING/ALLOCATING
INVOCATION:    IF WFM USED
INFO REQUIRED:  SEE OR-10
                U = WI-IMPLICATED DATA RESIDENCY ON TIER(DAL) (MB)
                V = WI-IMPLICATED DATA RESIDENCY ON TIER(X) (MB)
RULE:  MU>>U(TIER)=(2*V), ELSE MD>>U(TIER)=(2*V)
*************************************************************
FACTOR:  OR-12  WEIGHT:  5  TITLE:  WI FUNCTION -SEARCHING/
                                    REFERENCING
```

INVOCATION: IF WFM USED
INFO REQUIRED: SEE OR-10, OR-11
RULE: SEE OR-11

****

FACTOR: OR-13 WEIGHT: 5 TITLE: WI FUNCTION - READING/REVIEWING
INVOCATION: IF WFM USED
INFO REQUIRED: SEE OR-10, OR-11
RULE: SEE OR-11

****

FACTOR: OR-14 WEIGHT: 5 TITLE: WI FUNCTION - FILING/
DISPATCHING/MESSAGING
INVOCATION: IF WFM USED
INFO REQUIRED: SEE OR-10, OR-11
RULE: SEE OR-11

****

FACTOR: OR-15 WEIGHT: 5 TITLE: WI FUNCTION - TYPING/
ENTERING/INPUTTING
INVOCATION: IF WFM USED
INFO REQUIRED: SEE OR-10, OR-11
RULE: SEE OR-11

****

FACTOR: OR-16 WEIGHT: 5 TITLE: WI FUNCTION - COMPUTING/
CALCULATING/PROCESSING
INVOCATION: IF WFM USED
INFO REQUIRED: SEE OR-10, OR-11
RULE: SEE OR-10 AND OR-11
VERIFY THAT ANY DOWN-TIER MOVEMENT UNDER RULE OR-11 DOES NOT
CORRUPT TS-TO-TR RELATIONSHIPS.
SPECIAL CASE: ACQMDL=PLATFORM, WHERE C=1.00

****

FACTOR: OR-17 WEIGHT: 5 TITLE: WI FUNCTION - TELECOMMUNICATING
INVOCATION: IF WFM USED
INFO REQUIRED: SEE OR-10, OR-11
RULE: SEE OR-11

****

FACTOR: OR-18 WEIGHT: 5 TITLE: WI FUNCTION - MEETING
INVOCATION: IF WFM USED
INFO REQUIRED: SEE OR-10, OR-11
RULE: SEE OR-11

****

FACTOR: OR-19 WEIGHT: 5 TITLE: WI WORKPLACE SPECIFICITY

INVOCATION: IF WFM USED

INFO REQUIRED: K = TIER VALUE OF WORKPLACE MOST CLOSELY LOCATED TO WI CLASS N

L = NUMBER OF WORK ITEMS IN WI CLASS N

T = NUMBER OF WORK ITEM CLASSES IN APPLICATION

A = SIGMA (L(N)*K(N), FOR N=1 . . T)

B = SIGMA (L(N), FOR N=1 . . T)

C = A/B

TIER (Z) = ROUNDED VALUE OF C

RULE: E=TIER(Z)-TIER(DAL)

EXAMPLE:

CLASS	# IN CLASS	TIER	A VALUE	B VALUE
WI-1	500	1	500	500
WI-2	300	2	600	300
WI-3	200	2	400	200
		TOTAL	1500	1000

THEREFORE, C=1.5 AND TIER(Z)=2

****

FACTOR: OR-20 WEIGHT: 5 TITLE: WORKGROUP SECURITY REQUIREMENT

INFO REQUIRED: A = WG SECURITY REQUIREMENT (PER NCSC)

B = SYSTEM RATED SECURITY LEVEL (PER NCSC)

RULE: MU>>B=A OR MU>>B>A

****

FACTOR: OR-21 WEIGHT: 5 TITLE: USE OF SWI MODEL

INVOCATION: IF SWI MODEL IS USED

RULE: IF TIER(DAL)=1, E=1, ELSE E=0

****

FACTOR: OR-22 WEIGHT: 5 TITLE: EXTERNAL WI TIER GRAVITATION

INVOCATION: IF WFM USED

INFO REQUIRED: A = WI(A)

X = WI(X) (WIs WITH EXTERNAL RELATIONSHIP)

(ABOVE TWO MEASURES ARE SIGMA FOR ALL CLASSES)

RULE: MU>>X<(.20*A), ELSE MD>>X<(.20*A)

****

FACTOR: OR-23 WEIGHT: 5 TITLE: MANAGEMENT OF INFORMATION AS A CORPORATE RESOURCE

INVOCATION: ALL CASES

INFO REQUIRED: A = IM LEVEL OF TIER(DAL)

B = IM LEVEL REQUIRED BY APPLICATION AND/OR BY IT FUNCTIONAL RULING

RULE: MU>>A=B

```
**************************************************
FACTOR:  ER-1   WEIGHT:  4  TITLE:  TECHNOLOGY RATIO
INVOCATION:    ALL CASES
INFO REQUIRED:  TS(A), TS(T) PER SFM
                A = TS(A)/TS(T)
RULE:  MU>>A<.50
COMMENT:  RULE ADDRESSES THE PORTION OF TOTAL TS CAPABILITY DEDICATED
          TO THE APPLICATION ON TARGET PLATFORM
**************************************************
FACTOR:  ER-2   WEIGHT:  4  TITLE:  ATTRIBUTABLE COMMUNICATIONS
                                    COSTS
INVOCATION:    WHERE APPLICATION AND DATA CAN BOTH BE IMPACTED BY THE
               PLACEMENT DECISION
INFO REQUIRED:  TIER(C) = TIER OF RESIDENCY OF ENTITY TO BE
                COMMUNICATED WITH (USUALLY THIS IS ANOTHER
                APPLICATION OR A PACKAGE)
RULE:  !MIN((APPL-ENTITY COMM. COST)+(APPL-USER COMM. COST)) TIER 4-
       TO-1 (-3) TIER DISPLACEMENT DISALLOWED
COMMENT:  NOTE PREFERENCE ORDER:
               A - SAME SYSTEM -> SAME TIER (1..3)
               B - SAME LAN———> SAME OR CONTIGUOUS (1,2)
               C - SAME FACILITY-> SAME OR CONTIGUOUS (1,2)
               D - SAME AREA——> (1..3)
               E - DIFFERENT AREA-> (1..4)
          IN GENERAL, UP-TIER MOVEMENT RESULTS IN PLATFORM
          CONSOLIDATION AND DOWN-TIER MOVEMENT RESULTS IN PLATFORM
          REPLICATION.
**************************************************
FACTOR:  ER-3   WEIGHT:  4  TITLE:  APPLICATION PRODUCT PRICING
                                    POLICY
INVOCATION:     ALL CASES
INFO REQUIRED:  LRU = LICENCE RATED USER
                AU = ACTUAL USER
                CST = COST OF APPLICATION
                P(TIER) = CST(TIER)/((LRU+AU)/2)
RULE:  MU>>P(TIER)=MIN, ELSE
       MD>>P(TIER)=MIN
**************************************************
FACTOR:  ER-4   WEIGHT:  4 TITLE:   COST PER SFM TS,CS AND SS POINT
INVOCATION:    ALL CASES
INFO REQUIRED:  TS(A) = TS ALLOCATED TO APPLICATION
                SIMILARLY, CS(A) AND SS(A)
```

```
RULE:  MU>>!MIN(COST/TS(A)+COST/CS(A)+COST/SS(A)), ELSE
       MD>>!MIN(COST/TS(A)+COST/CS(A)+COST/SS(A))
**************************************************
FACTOR:  ER-5   WEIGHT:  4 TITLE:   APPLICATION SPECIAL CASE PRICING
INVOCATION:    SUPPLIER MAKES 'SPECIAL OFFER'
INFO REQUIRED:  SEE ER-3
RULE:  RE-ACTIVATES FACTOR ER-3 ANALYSIS
**************************************************
FACTOR:  ER-6   WEIGHT:  4 TITLE:   IMPLEMENTATION AND SUPPORT COSTS
INVOCATION:    ALL CASES
INFO REQUIRED:  A = LIFE CYCLE SUPPORT COST/INTENDED USER
RULE:  CONSISTENT WITH APPL REQUIREMENTS, MD>>A=MIN, ELSE MU>>A=MIN
**************************************************
FACTOR:  TM-1   WEIGHT:  3  TITLE:  NETWORK SERVICE REQUIREMENT
INVOCATION:    APPL IS TO ACCESS NETWORK
INFO REQUIRED:  A = WIs IN CLASS A (# OF WIs)
                B = LOGIC INTENSITY (1 . . 3)
                C = DATA (MB/10)
                D = TRANSACTION INTENSITY (1 . . 3)
RULE:  MU>>SIGMA(A*B*C*D)=MIN, ELSE
       MD>>SIGMA(A*B*C*D)=MIN
**************************************************
FACTOR:  TM-2   WEIGHT:  3  TITLE:  APPLICATION DEGREE OF
                                    COMBINABILITY
INVOCATION:    GENERIC SOURCE = THIRD PARTY; AND
               APPLICATION IS TIER-LIMITED
INFO REQUIRED:  TIER(TP) = TIER TO WHICH APPLICATION IS LIMITED
RULE:  E=TIER(TP)-TIER(DAL) (RUN RULE ITERATIVELY AND CAPTURE
       MULTIPLE SCORES IF APPLICATION RUNS ON LESS THAN FOUR TIERS.)
**************************************************
FACTOR:  TM-3   WEIGHT:  3  TITLE:  PROJECT TEAM BIAS
INVOCATION:    ANALYST
RULE:  SUBJECTIVE IN FAVOUR OF TIER(X)
       E=TIER(X)-TIER(DAL)
**************************************************
FACTOR:  OM-1   WEIGHT:  2  TITLE:  PROJECT TEAM BIAS
INVOCATION:    ANALYST
INFO REQUIRED:  X = TOTAL EST. LIFE CYCLE NPV OFFSET IN FAVOUR
                    OF TIER(Y)
                A = COST(TIER Y) - OFFSET
                N = TIER VALUE OF TIER UNDER CONSIDERATION
```

```
RULE:  WHERE A>COST(N) FOR ALL VALUES OF N FOR OTHER TIERS,
       DECLARE TIER (Y) TO HAVE A COST ADVANTAGE
       E=TIER(Y)-TIER(DAL)
```

Appendix C: PIMOCS data set record structures

This appendix provides information on the major data sets used within PIMOCS and has been designed to support both manual and automated usage.

Each data set is defined in terms of the record structure, incorporating fields and (sample or full universe) values for those fields.

```
*****************************************************
WI            Work Item Data

CLASS:        (Work Item Class)
NUMBER:       (Number of WIs in this class)
RELATED:      (Related classes of work items)
NAME:         (Text name of WI)
ORIGIN:       (Name of origin workgroup of WI)
DESTN:        (Name of destination workgroup of WI)
WPORIG:       (Identifier of origin workplace of WI)
WP2 . . WP8   (Subsequent workplace identifiers within WI journey)
WPDEST:       (Identifier of destination workplace)
PRIORITY:     (One of seven levels below)
                   -   URGENT/INTERRUPT
                   -   EXTERNAL
                   -   CONDITION-GENERATED
                   -   BRING FORWARD
                   -   ONGOING
                   -   ROUTINE
                   -   RESIDUAL
TH:           (Time required in the abstract thinking move to complete
              this work item, expressed in hours - ex: 2.80 HRS)
              Similarly, required time to complete any/all of the
              following:
EV            Evaluating
AN            Analysing
CO            Comparing/contrasting/sorting/allocating
SR            Searching/retrieving
RR            Reading/reviewing
FM            Filing/dispatching/messaging
TY            Typing/entering/inputting
CP            Computing/calculating/logically processing
TL            Telecommunicating
```

TR Travelling
CM Composing
(NOTE: The use of WFM within PIMOCS is optional.)

****

COMPONENT System Architecture Component Definitions

Basic Information

GENERIC NAME: DATABASE SYSTEM FOR MRS (50 USER) ACRONYM: RDBMS
VENDOR: DESCRIPTION: RDBMS ENGINE PRODUCT SET
SUPPORTS: APPLICATION
PEER1: OA PKG CLASS: PACKAGE PEER2: LANG/APPL
REQUIRES: SYSTEM
SUPPLY MODE: MIXED

Cost Information

CAPITAL
Corporate Planning/Acquisition/Management .... $
Purchase/Delivery 50000
Installation/System Integration/Commissioning ....
Orientation/Training.(Admin/User) 20000
TOTAL CAPITAL .... $ 70000
OPERATION AND MAINTENANCE
Corporate Support .... $
Annual Licence Fees 3000
Lease/Rental ....
Software Support/Upgrade 3000
Hardware Maintenance ....
Administrator Salary and Related O&M ....
Administration or Facility Management Contract ....
Supplies and Expendables ....
Re-location/Re-installation ....
Electrical ....
Dedicated facility or Support Equipment ....
Other ( ) ....
Other ( ) ....
TOTAL O&M .... $ 6000

Key Attribute/Characteristic Value Summary

ATTRIBUTE	VALUE	TS	CS	SS
RDBMS ENGINE	50 USER	150000		
DOCUMENTATION	USER			50000
CBI	USER			50000
VENDOR SUPPORT	USER			50000
TRAINING	USER			50000

Component Stack Profiles

<table>
<tr><th colspan="3">TECHNOLOGY
(TS)</th><th>COMMUNICATIONS
(CS)</th><th>SUPPORT
(SS)</th></tr>
<tr><td rowspan="3">P
A
C
K
A
G
E
( )</td><td>APPL.
( )</td><td rowspan="2">APPL.
( )</td><td>APPLICATION
( )</td><td>SELF
(50000)</td></tr>
<tr><td rowspan="2">PKG.
(150000)</td><td>PRESENTATION
( )</td><td>SPECIALIST
( )</td></tr>
<tr><td>LANG.
( )</td><td>SESSION
( )</td><td>LANA/MRSA
( )</td></tr>
<tr><td colspan="3" rowspan="2">OPERATING
SYSTEM
( )</td><td>TRANSPORT
( )</td><td>USER-ASSIST
( )</td></tr>
<tr><td>NETWORK
( )</td><td>VENDOR
(50000)</td></tr>
<tr><td colspan="3" rowspan="2">HARDWARE
( )</td><td>DATA LINK
( )</td><td>CBI
(50000)</td></tr>
<tr><td>PHYSICAL
( )</td><td>DOCUMENTATION
(50000)</td></tr>
</table>

****

CONFIG Candidate Application/System Combination

Basic Information

GENERIC NAME: MIDRANGE SYSTEM (50 USER) ACRONYM: MRS/50

VENDOR: DESCRIPTION:

SUPPORTS: LANG, PKG, APPL

PEER1: MRS/20 CLASS: SYSTEM PEER2: MRS/30

REQUIRES:

SUPPLY MODE: MIXED

Cost Information

CAPITAL

Corporate Planning/Acquisition/Management .... $ 20000
Application 50000
Purchase/Delivery 140000
Installation/System Integration/Commissioning 15000
Orientation/Training 25000
TOTAL CAPITAL $250000

OPERATION AND MAINTENANCE

Corporate Support ....$ 10000
Annual Licence Fees ....
Lease/Rental ....
Software Support/Upgrade 2000
Hardware Maintenance ....
Administrator Salary and Related O&M 40000
Administration or Facility Management Contract ....
Supplies and Expendables 5000

Re-location/Re-installation 2500
Electrical 500
Dedicated facility or Support Equipment 500
Other (   ) 2000
Other (   ) ....
TOTAL O&M .... $ 62500

`*************************************************`

Key Attribute/Characteristic Value Summary

ATTRIBUTE	VALUE	TS	CS	SS
CPU	RISC/25–30 MHz	25000		
CPU	(SAME)	25000		
MEMORY	32 MB	3500		
AUX. MEMORY	32 MB	3500		
STORAGE	1000 MB	10000		
EXT. STORAGE	2000 MB	20000		
FP PROCESSOR		5000		
PRINTER (LSR)	2	6000		
PRINTER (PIN)	2	2000		
SNA EMUL.	10 USER		5000	
X.25 EMUL.	30 USER		15000	
OS	UNIX	320000		
LANG	C/COBOL	50000		
DOCUMENTATION	USER/REF/2 WAN			100000
CBI	UNIX WAN			100000
MRSA				100000
VENDOR SUPT.	UNIX WAN			100000
TRAINING	UNIX WAN			100000

Component Stack Profiles

<table>
<tr><th colspan="3">TECHNOLOGY
(TS)</th><th>COMMUNICATIONS
(CS)</th><th>SUPPORT
(SS)</th></tr>
<tr><td rowspan="3">P
A
C
K
A
G
E
(   )</td><td>APPL.
(   )</td><td rowspan="2">APPL.
(30000)</td><td>APPLICATION
(20000)</td><td>SELF
(50000)</td></tr>
<tr><td rowspan="2">PKG.
(   )</td><td>PRESENTATION
(20000)</td><td>SPECIALIST
(   )</td></tr>
<tr><td>LANG.
(50000)</td><td>SESSION
(20000)</td><td>LANA/MRSA
(100000)</td></tr>
<tr><td colspan="3" rowspan="2">OPERATING
SYSTEM
(320000)</td><td>TRANSPORT
(20000)</td><td>USER-ASSIST
(   )</td></tr>
<tr><td>NETWORK
(20000)</td><td>VENDOR
(10000)</td></tr>
<tr><td colspan="3" rowspan="2">HARDWARE
(100000)</td><td>DATA LINK
(20000)</td><td>CBI
(10000)</td></tr>
<tr><td>PHYSICAL
(20000)</td><td>DOCUMENTATION
(10000)</td></tr>
</table>

NOTE: Training on networks is for ALL users who contend for the network access positions.

****
COB Cost-Opportunity-Benefit Records

NATURE: (STR = Strategic, TAC = Tactical)
TYPE: (COST, OPP, BEN)
CODE: (Reserved for future use)
DESIGNATOR: (Here, a number issued to identify an application candidate and is elsewhere tied to an APPL database record while a letter is used for a system-only candidate and an alpha-numeric combination is used for a combination application/system candidate, each of which is a CONFIG record - for example:
'2' is an application
'B' is a system
'B2' is an application/system combination
NAME: (Text name of this COB item)
DEFN: (Logical or other definition of COB item)
PROFILE: (This is a sub-table indicating any required breakdown of components within this record - for example, an on-site service/support contract might have various elements which vary in cost from year to year) (For each component, costs and quantifiable negative results of opportunities are shown as negative real numbers while benefits and positive opportunity results are shown as positive real numbers)

ITEM	YEAR 0	YEAR 1	YEAR 2	YEAR 3	YEAR 4	YEAR 5
NET						

TOTAL: (Sum of all net amounts from table, expressed as a real number)
AVG: (Average of net amounts)
NPV: (Net Present Value of net amounts)
****
FACTOR Application Placement Factor Rules

(See Appendix B)
****
PLCMT Application Displacement Findings

APPL: (Application designator, as above)
TYPE: (Type of finding: ABS, REL or MIN)
FACTOR CODE: (Code of factor driving this placement finding)
FACTOR NAME: (Name of factor driving this placement finding)
DIRECT FINDING: (Displacement action from finding - ex: ''+3'' represents a vote to move up three tiers)
INDIRECT FINDING: (Tier we will arrive on if we implement the DIRECT FINDING - in this case, if the default tier is Tier 1, this finding would move us to Tier 4)

`*****************************************************`

WG Workgroup Environment Data

NAME: (Workgroup text name)
FUNCTION: (Text description of workgroup function)
PROFILE: (This is a sub-table profiling the number of users in each class as below)

USER CLASS	NUMBER	TR TOTAL	CR TOTAL	SR TOTAL
A	20	20000	0	20000
B	10	30000	20000	30000
ALL	30	50000	20000	50000

SOPH: (Degree of workgroup sophistication with respect to systems technology - a possible scale of 0–10 with 10 as the highest experience level is recommended)
COMMENT: (Text commentary on workgroup requirements)
REGISTRY: (Software Registry Application Profiles)
REGCODE: (Registry's own code assigned to the application)
TYPE: (Application type code assigned by Registry)
SITES: (Number of sites or platforms running the application inside the organization, division etc.)
NAME: (Text name of application)
ORIGIN: (Text name of origin organization)
IOC: (Date of initial operational capability in the organization)
ORIGINAL COST: (Cost to buy or develop to IOC)
VALUE: (Total investment in application by organization)
REQUIRES: (Codes of system architecture components required by this application - ex: RDBMS, MRS50 etc.)
REQUIRED BY: (Codes of any system architecture components which themselves require this application)
FUNCTIONS: (Text description of application subject coverage and functions arranged for keyword searching)
DEPLOYMENT: (Sub-table indicating the deployment status of the application - users and the as-installed capacity)

VERSION	DESCRIPTION	SITE	USERS	TS	CS	SS
1.15	UNIX MRS	SALES	25			
2.01	UNIX MRS	PROD.	11			

(In the above table, the example TS, CS and SS scores would connote the total capability of the application. If it was rated for 50 users and provided a basic capability of TS=1000, then the TS score for each of the Sales and Production Departments would be 50000 even though they only had 25 and 11 users respectively.)
BASE: (Totals of TS, CS, and SS installed capacity)
COMMENT: (Text comment) CONTACT: (Application manager in Registry)

`*****************************************************`

APPL Candidate Application Profiles

REGCODE: (Registry code, as above)
DESIGNATOR: (Candidate number or one of the following:
CU = Current Application
RQ = Required Application)
NAME: (Text name of Application)
ACRONYM: (Acronym representing application)
DESC: (Text description of application)
TR: (Technology requirement of application and/or system, as imposed on the lower layers of the TR stack)
CR: (Communications services requirement, as imposed on the lower layers of the CR (OSI) stack)
SR: (Support requirement as imposed on the workgroup's own internal support resources (ex: system administrator), IT support and the vendor)
TS: (Technology capability provided by the application/system)
CS: (Communications services provided)
SS: (Support services provided)
PROFILE: (Sub-table of component parts of application indicating their cost profiles)

ITEM	YR 0	YR 1	YR 2	YR 3	YR 4	YR 5
BUY	125000					
LICENCE		5000	5000	5000	5000	5000
SUPPORT		15000	15000	15000	15000	15000

(These cost items become negative amounts within COB records)
COMMENT: (Text commentary)
CONTACT: (Registry contact person)
SVCEMDL: (Service model S-U, M-U or C-S)
USERS: (Nominal number of users which can be hosted by the application, where necessary segmented by any applicable platform-induced limitations at various tiers)
COST: (Cost to buy, modify to develop to IOC point)
REQUIRES: (Codes of system architecture components required by this application – ex: RDBMS, MRS50 etc.)
REQUIRED BY: (Codes of any system architecture components which themselves require this application)
FUNCTIONS: (Text description of application subject coverage and functions arranged for keyword searching)
GENSOURCE: (Generic Source: use one of:
REGISTRY
MODIFY
THIRD PARTY
DEVELOP)
SPECSOURCE: (Specific Source: text name of supplier)
TRADENAME: (Trade name of application if different from ACRONYM)
VER: (Version under consideration)
REL: (Release under consideration)

****

ACQ Acquisition Procedures

GENERAL: (One of the following:

BUY Purchase outright
LEASE Lease application/system
RENT Short-term rental only
BORROW Obtain loan of system to workgroup
TRIAL Vendor/IT Group joint funded trial
PILOT Vendor/IT/user funded production pilot
FREE No-cost item)

ASSET: (System architecture component code)
TYPE: (Generic text description of asset)
NAME: (Text description of item to be acquired)
PROFILE: Sub-table indicating up to ten steps in free form text)
INVOKED: (Conditions invoking this type of acquisition procedure defined logically in terms of entities, relationships and processes known to PIMOCS)
EXCLUDED: (Conditions barring this procedure)
TIME: (normal or average elapsed time to implement this procedure)
COST: (Cost profile of any overhead acquisition costs. This is usually worth profiling only in very large private firms or in a governmental organization.)

ITEM	YR 0	YR 1	YR 2	YR 3	YR 4	YR 5
RFP	10000					
BIDDING	5000					
AWARD		2500				

****

VENDOR Vendor Product/Certification Database

CODE: (Vendor identification code)
NAME: (Text name of vendor)
CLASS: (System architecture component code)_
PRODUCT: (Text name of product)
PROFILE: (Sub-table of components and certification status)

VENDOR	PRODUCT	CATEGORY	STATUS
ORACLE	V.7 FOR UNIX	RDBMS	APPROVED
INGRES	PENDING	RDBMS	PENDING

CONTACTS: (Sub-table of key vendor contacts)

SENIOR Senior Management
MARKETING Marketing
HARDWARE Hardware specialists
SOFTWARE Software specialists
COMM Communications specialists
SUPPORT Support Organization
CONTRACTS Contract Administrator

****

OBJEC Project Objectives

NUMBER: (Project Objective Number, to be arranged in priority order)
NAME: (Text name of objective)
DESC: (Full text description of objective)
TRXL: (Translation of objective, where possible, into a logical arrangement of entities, relationships and processes which is intelligible to PIMOCS ex: if an objective is that there be no more than 50 users on the system, the logical expression "USER.LE.50")
COBREF: (COB dataset cross-reference, where applicable)
PROFILE: (Where the objective concerns cost containment or opportunity exploitation - for quantitatively determinable expected amounts of positive impact or similarly quantifiable benefits, this sub-table provides a full profile)

ITEM	YR 0	YR 1	YR 2	YR 3	YR 4	YR 5
CAPITAL (C)	-250	-50	-25	-10	-10	-10
OPERATING (C)		-100	-100	-100	-100	-100
CUT SPACE (O)		25	50	75	75	75
NEW SALES (O)		50	75	100	100	100
ACCESS (B)		5	10	10	10	10
OA BEN. (B)		100	100	100	100	100
LESS TRG. (B)			5	20	30	10
NET	-250	30	145	195	205	185

The above table, for example, shows capital and operating outflows but compensating positive opportunities such as cutting space due to workplace distribution (including to the home workplace in some cases) and the opportunity to increase sales due to higher salesforce productivity because of the new system. The pure benefits here are from better mainframe access, office automation and less re-training. All figures represent thousands of dollars. Here, ignoring cost of capital, payback is accomplished during the third year. This data is compatible with the COB data set.

Appendix D: PIMOCS module definitions

This appendix contains a manual specification, in purely functional terms, of each functional module of the planning and implementation model of open computing systems (PIMOCS). It provides a mostly text-algorithmic and general treatment of each PIMOCS function. Variable names, where given, are only intended as recommendations. Note that in many cases variable names are described in text only and are not fully quantified or limited in range, scope, storage format, presentation etc. Some variables may cross-reference to (and in the automated version benefit from 'hot links' to) fields within various data set record structures. Where all or part of PIMOCS is automated, it is strongly recommended that as many variables as possible be implemented as RDBMS table positions (i.e. as fields), even when they are not arrays. This will permit easier expansion and thus will more readily facilitate customization in the future. From a design perspective, it may be desirable to group all variables related to system architecture and system environment into a single data set. No decisions have been taken in defining PIMOCS which would preclude this; however, such a step is viewed as optional. The amount of functionality embodied in most of the defined PIMOCS modules is quite small; therefore a large number of modules has been defined. This permits ready modularization of code for PIMOCS and should also assist in trouble-shooting a final design or actual code. Conversely, individual module functional (and also non-database variable) definitions are quite wide, permitting the designer to:

- structure data according to the preferred design approach or according to the dictates of a CASE tool as required
- perceive and exploit any interrelationships among data elements (as captured in fields, variables and arrays) which are not explicitly treated at this stage, but which would add flexibility to the implementation of PIMOCS.

The fully qualified naming of modules shall be in the form of:

```
XX/YYY/ZZZZ
```

where:

XX = Phase, being one of:

- IN Project initiation
- CL Classification
- RQ Requirements
- IV Investment objectives
- FS Feasibility
- BC Business case

YYY = Stream, being one of:

- EXEC Executive
- SFM System Functional Model
- APM Application Placement Methodology
- EMIT Economic Model of Information Technology
- OTHR Other items not covered above

ZZZZ = Module identifier code, which is unique to each module but will in many cases be built up from the following abbreviations:

- ACQ Acquisition
- APPR Approach
- ARCH Architecture
- ASSUM Assumptions
- BEN Benefits
- CAP Capital approval document
- COST Costs
- DAL Default application locus
- DEPL Deployment
- ENVT Environment
- ESTIM Estimate
- EXT External
- GEN General
- IMPL Implementation
- INFO Information
- OPP Opportunity
- PLG Plan/planning
- PRC Procurement review council or committee
- PROC Procurement
- REQ Requirement
- SIT Situation
- STR Strategic
- SVCE Service
- SYS System
- TAC Tactical
- TAXN Taxonomy

USR	User
WFM	Workflow methodology
WG	Workgroup
WP	Workplace
WI	Work item

The PIMOCS functional module definitions in this appendix are presented in the following format, which has been set up to support both a workbook approach and the development of an automated tool specific to your organization's requirements.

CODE: (Fully qualified module name, as defined above)
NAME: (Text name of functional module)
FUNCTIONS: (Descriptions of function of module in text and/or in logical expressions relating to entities, processes and relationships known to PIMOCS)

```
CODE: IN/SFM/USRENVT    NAME: User Environment Information
1. Obtains and classifies TR, CR and SR profiles of individual
workers who make up the WG which will use the application and/or
system.
2. Stores information in WG dataset.
**************************************************
CODE: IN/APM/NOMREQ    NAME: Nominal Requirement
1. Accept free-style text input and/or TR/CR/SR descriptions of the
WG's application requirement based on initial information available
to the analyst.
**************************************************
CODE: IN/APM/WGENVT    NAME: Workgroup Environment/Assumptions
1. Capture and record information about the workgroup and its
environment as set out in the APM. Such constraints, to be expressed in
logical terms impacting elements, relationships and processes known
to PIMOCS. Key items include budgetary constraints, political
economy of project doability, management degree of comprehension of
the requirement, end-user sophistication (1-10 scale per WG
sophistication), security requirement (per NCSC), conflict of user
desires and principles of good system management, stability of
workflow (variability %) and deployment strategy.
**************************************************
CODE: IN/EMIT/PROJ    NAME: Project Initiation Data
1. Establish project boilerplate data per the organization's
current SDLC document.
```

****

CODE: IN/EMIT/ECOENVT NAME: Economic and Budgetary Environment
1. How many people are in the workgroup? What size of application and/ or system is contemplated? What order of magnitude of benefits and costs are expected?

****

CODE: IN/OTHER/CONSTR NAME: Constraints Identification
1. Identify and define constraints in terms of system architecture components, processes and relationships known to PIMOCS with a lexicon of logical operands, plus variables selected from a menu. Ex: TIER(DAL).NE.2 or SVCEMDL.EQ.MU

****

CODE: IN/OTHER/CHTR NAME: Project Charter
1. Finalize all project charter boilerplate per SDLC.

****

CODE: CL/SFM/PROGPLAT NAME: Select Acquisition Model
1. Select PROGRAM model where:
- OA is included
- there are two or more applications to be run
- either condition is anticipated for the future.
2. Select PLATFORM model where:
- there is one application only
- there are two applications now but none later
- system is operational and/or embedded in nature.

****

CODE: CL/SFM/COMPONENT NAME: Identify Components for Analysis
1. Identify which system architecture components will be the subject of the analysis from among: application only, application/system combination, system only and other/unknown.

****

CODE: CL/SFM/SVCEMDL NAME: Service Model
1. Permit early pro-forma identification of service model from among SU, CS or MU and store result as a variable, in an APPL and/or CONFIG record or COMPONENT record if such exists. OR
2. Invoke FS/APM/SVCEMDL2 routine to permit full determination of service model by rule-based method.

****

CODE: CL/SFM/SYSARCH NAME: System Architecture Assumptions
1. Import set of default system architecture assumptions drawn from the APM. These must support the assumptions made later in support of DAL. Specifically, they are: ARCHAPPR=UNIFIED, FA=OPERATIVE,

TIERS=4, SFM=ACTIVE, SFM CALIBRATION CONFIDENCE = HIGH, SVCEMDL=PROGRAM OR PLATFORM, OS=UNIX, Pillars of Portability and Continuum of Portability models in use and DAL assumptions are operative.
2. Additional logical constructions would also be allowable. For example, if MBS is to be disallowed for WGs of over ten users, it could be expressed thus: WHERE WG>10, SYS.NE.MBS

****

CODE: CL/APM/TIEREXP NAME: Expected or Known Tier
1. Where the WG and/or project team have a pre-conception or foreknowledge as to likely or certain tier selection outcome, it can be declared here.

****

CODE: CL/APM/DEPLSTR NAME: Deployment Strategy
1. Obtain and store information on the a priori determination, if any, of deployment strategy among the following:

1 TRIAL —> PRODUCTION
2 TRIAL –> PILOT –> PRODUCTION
3 PILOT –> PRODUCTION
4 LAB –> TRIAL –> PRODUCTION
5 LAB –> TRIAL –> PILOT –> PRODUCTION
6 LAB –> PILOT –> PRODUCTION

****

CODE: CL/APM/USRTAXN NAME: User Taxonomy
1. User groups can be classified by such items as: degree of sophistication, current system environment, role or function in relation to the application itself and resource requirements (TR, CR and SR per SFM).
2. In setting a unit or workgroup's User Demand Profile (UDP), per the SFM, use a THRESHOLD APPROACH. Each user must be accounted at his or her most demanding use.

****

CODE: CL/EMIT/STNDG NAME: Project Standing
1 Select one from the following:
- DE NOVO (STARTUP)
- RE-START
- UPGRADE/EXTEND EXISTING SYSTEM

****

CODE: CL/EMIT/PLGENVT NAME: Planning Environment
1. Accept and store information for the following, as per EMIT: deployment mode (REPLICATIVE, RELATIVE or ABSOLUTE), inflation, tax

application, wage increase factor, discount rate for COBA, project life (default is 5 yrs) and profiles of PRIME, SECONDARY and TERTIARY periods.

****

CODE: CL/EMIT/SYSENVT NAME: System Environment Assumptions

1. Offer default set of system assumptions as per the APM in Chapter 5 (basically, an early version of the DAL assumption set) for review/ alteration by user. If prompted, branch to BUILD to construct a model of the WG's existing system in SFM terms.

****

CODE: CL/EMIT/MOTIVE NAME: Project Motivators

1. Prompt user to offer assistance in creation of the first few OBJECT dataset records with one record each for the following: maximize workgroup efficiency, obtain available application, automate manual process, accept application offered/mandated by HQ Group Head, take advantage of latest available technology to replace old, obsolete or failing system.

****

CODE: CL/EMIT/ANALTYPE NAME: Select Type of Analysis

1. Select conventional analysis or WFM for use in the requirements phase.

****

CODE: CL/OTHR/KILL NAME: Terminate Project for Cause

1. Enter text narrative regarding cause of termination. This should also invoke and fill an OBJECT record. Offer user the opportunity to re-visit any/all datasets and alter fields manually.

****

CODE: CL/OTHR/PROSPECT NAME: Finalize/Approve Project Prospectus

1. Collect and modify summary information by plucking it from the WI, WG and APPL datasets and from variables added in the routines listed below. Relevant modules are: MOTIVE, NOMREQ, WGENVT, USRENVT, USRTAXN, SYSENVT, ECOENVT, PLGENVT, PROJ, STNDG, SYSARCH, SVCEMDL, ANALTYPE, PROGPLAT, SVCEMDL2 and DEPLSTR.

****

CODE: RQ/SFM/REQTRXL NAME: Requirements Translation

1. The purpose of this module is to relate stated user requirements as compared to known 'benchmarked' applications of quantified and understood levels of complexity (and hence TR/CR/SR capacity demand) so as to derive an approximate technology, communications and support requirement profile.

****

CODE: RQ/SFM/TR NAME: Technology Requirement

1. Set Level 8 (user) TR score; OR
2. Build top-down profile by, for example, invoking ORACLE for 30 users and (where PIMOCS is automated) permitting PIMOCS, from its knowledge of the intended system and ORACLE, to fill in the lower layers or else offer a choice of various alternative possible configurations, each in an automatically generated CONFIG record. In a manual run situation, use your previous assignment of TR and TS ratings to support this buildup.

****

CODE: RQ/SFM/CR NAME: Communications Requirement

1. Set Level 8 (user) CR score; OR
2. Build top-down profile by, for example, invoking a communications requirement stated in terms of elements, relationships and processes and then fill in the lower layers or else offer a choice of various alternative possible communication service combinations, each in a CONFIG record.

****

CODE: RQ/SFM/SR NAME: Support Requirement

1. Set Level 8 (user) SR score.

****

CODE: RQ/APM/WORKENV
NAME: Work and Work Flow

1. Where WFM is used, construct a 'T' diagram with work flow 'overhang' (over the triple stack) based on the percentage of the workgroup's WIs which are performed in the current business process (however handled) and in the proposed application. This provides an indication of what percentage of the workgroup's current business will be addressed by the new application.

****

CODE: RQ/APM/APPLTAXN
NAME: Application Taxonomy

1. Enter or modify APPL dataset as required to characterize the desired application. This record or records of APPL dataset are of type RQ and do not have application numbers assigned unless:
   - the requirements is stated in terms of the existing (as-installed) application; and/or
   - such application is also to be treated as one of the candidates.
2. Collect information from variables loaded in previous routines respecting all of the following: DAL, SVCEMDL, NMREQ and APPLASSUM.

(NOTE: Normally, for a given PIMOCS analysis, there will be several records compiled under APPL, including:
- current application .... CU
- application requirement .... RQ
- one for each defined and declared candidate application.... 01-99
- current application where it is a candidate 1R

****

CODE: RQ/EMIT/SUPTENVT NAME: Support Environment Assumptions
1. Enter information regarding the support model, and in particular information regarding any first year and subsequent-year levies from corporate IT to the WG connected with the various options of application, system and application/system combinations which have been identified.

****

CODE: RQ/OTHER/GENREQ NAME: Generic Requirements
1. Permit entry of additional requirements information in the analyst's choice of: Free form text OR Logical expressions defining existing or required relationships among entities and processes intelligible to PIMOCS.
2. Identify any logical controls or constraints on the application, system or project.
3. Identify any bilingual or multi-lingual requirement by creating a special CONFIG record indicating the degree of such requirement for each component of the triple-stack SFM model.
4. Identify any other project-impacting circumstances. Where all else fails, invoke MOTIVE and create a free-form OBJECT record.

****

CODE: RQ/OTHER/WFM NAME: Workflow Methodology
1. Implement Workflow Methodology/Model per Appendix A. The model flows WIs through workplaces within and outside the WG to simulate the performance of work both before and after the intended application and/or system is implemented.
2. Estimate the time savings potential due to any additional automation represented by the proposed application and/or system.

****

CODE: RQ/OTHER/REQSTMT NAME: Requirements Statement for Procurement Review Committee
1. Utilize the RQ record of the APPL dataset to generate a pro-forma requirement statement.

****

CODE: IV/EMIT/IMPLASSUM NAME: Application Envt. Assumptions
1. Provide the opportunity to re-access any of GENSOURCE, APPLTAXN, APPLASSUM or DAL if required.

****

CODE: IV/EMIT/INVCLX NAME: Investment Objective Classification
1. Review and complete OBJECT database.

****

CODE: IV/EMIT/OA NAME: Office Automation (OA) as a Special Case
1. Obtain information on the number of OA users in the WG, or in each WG if the application is to be replicated at multiple sites.
2. Generate a COB record, if one does not already exist, and for each intended user in the WG (or in each impacted WG) make a +10000 entry for each of the YR1-YR5 dollar benefits.
3. Set ACQMDL=PROGRAM and generate an 'OA' record within the OBJECT dataset if one does not already exist.

****

CODE: IV/OTHER/CAP1 NAME: Capital Approval Plan
1. Assemble, organize and produce information to be included in a preliminary capital plan or request document.

****

CODE: FS/SFM/APPLCHAR NAME: Application Characterization
1. Permit creation of a COMPONENT dataset record from a numbered APPL record (not CU or RQ but including 1R) for each specific application to be considered as a candidate.

****

CODE: FS/SFM/BUILD NAME: System Buildup
1. Assemble successive 'overlays' of the SFM display to establish a candidate system or application/system combination.
(NOTE: APPL record values of 01 . . 99 and CONFIG system record values of A . . Z are permitted. Here, of course, 'application' refers only to an end-user application, not a package such as ORACLE or an OA package. Where OA is the only application being considered, use the IV/EMIT/OA routine but change OA savings to zero if these are not considered as part of the justification statement.)
2. The system buildup routine, for adding a system architecture COMPONENT into the CONFIG being built, is as follows:

(1) Specify whether application/system or system only.
(2) Identify APPL and CONFIG input records which may already exist and fetch them.
(3) Consider Level 8 (user) TR/CR/SR profile.

(4) Select first/next generic component from listing derived from COMPONENT database, for addition to present CONFIG record. Optionally, also grant access to REGISTRY and APPL datasets to permit fetch of additional (possibly auxiliary) applications for addition to the CONFIG record.

(5) Where source is generic, select source-specific version of component to be added in.

(6) Display current information for the to-be-added COMPONENT, APPL or REGISTRY application, including: release, version, specifications and standards/profile adherence.

(7) Perform Three Axis Check for each level of each stack impacted by the addition. The three axis check addresses the following axes:

- VERTICAL for levels above/below in same stack
- HORIZONTAL for similar or analogous levels in one or two (as the case may be) contiguous stacks (only the Communications stack has two contiguous stacks)
- LONGITUDINAL for other overlayed components already in place

The check will be for capacity (VERTICAL downwards only), standards compatibility (all axis) and interoperability (all axis). Examples of constricting findings are set out below:

- hardware cannot support selected operating system
- specified communications network does not support the only protocol use by the hardware or RDBMS
- RDBMS requires a higher version of the operating system than has been selected
- selected RDBMS and OA packages have diametrically opposing GUI requirements.

(8) Perform SVCEMDL Implication Check for item just added to the overlay.

(A)

For SU, is the service model violated?

(B)

For CS, is the specified distribution of workload between client and server violated?

(C)

For MU, no checks required.

(9) Classify the built-up system as one of the following and take the appropriate action:

NEW SYSTEM - fetch cost info and create COB Cost record

UPGRADE - determine cost information

EXISTING - take no action

(10) Generate COB Cost record as decided at Step (9).
(11) Return to Step (4) for the next overlay, if any.
(12) END

****

CODE: FS/SFM/MATC NAME: Application/System Matchup
1. Display menu of application candidates, systems and combinations.
2. Permit matchup of applications and systems by permitting analyst to:
- select application and system
- observe full triple stack display
- re-invoke BUILD at user (or at PIMOCS) option when a conflict is detected - Ex: application running under INFORMIX is assigned to a platform shown equipped only with ORACLE

****

CODE: FS/APM/APPLASSUM NAME: Application Assumptions
1. Display and permit modification of assumptions relating to the application itself, which are common to all candidates except where indicated otherwise. Items to be addressed here include: SVCEMDL, SCALE, DAL, dependencies and requirements, any desired change to GENSOURCE and any desired change to SPECSOURCE.

****

CODE: FS/APM/SEQ NAME: Determine Feasibility Sequence
1. Refer to the rules set out earlier for determining whether to consider generic source first, specific source first or to consider all application issues before placement can influence the outcome of the APM.
(1) Signify as TRUE all valid rules;
(2) If no methods have TRUE rule conditions select Method A;
(3) If only one method has TRUE rule(s) select that method;
(4) If two or more methods have TRUE rules then total the numbers of the TRUE rules under each method and select the method with the lowest score (the rules are ranked in importance from greatest to least);
(5) If Step 4 produces a tie select Method A.

****

CODE: FS/APM/SVCEMDL2 NAME: Select Service Model
1. Refer to the rules set out in the APM which will be of assistance in selecting the service model for an application, where generic source or specific source candidate selection has not determined this already. Set SVCEMDL to SU, CS or MU.

****

CODE: FS/APM/DAL NAME: Establish Default Application Locus (DAL)

1. The DAL is established by re-setting any of the DEFAULT INDICATORS set out below, according to what is known about the application before the APM factors/rules are exercised. See Chapter 5, STEP 3 in APM procedure.
2. Store result under DAL variable.

****

CODE: FS/APM/CUSTOM NAME: Enter Custom Factors

1. Enter any additional application placement factors unique to this run of the APM.

  (1) IT and the client unit or workgroup identify what they agree is a valid APM factor, but one which is not included in the Stock universe of factors.

  (2) The IT analyst records important information about the proposed custom factor, to the greatest degree possible in terms of entities, functions and relationships already utilized within PIMOCS. Specifically, information for a custom factor would include:

  - factor identification / name
  - factor definition in terms of elements known and unknown to the system and the relationships among them
  - classification as primarily TECHNICAL, OPERATIONAL or ECONOMIC
  - classification as being ABSOLUTE, RELATIVE or MINIMUM in its impact upon tier selection
  - the RANGE of potential impacts or effects which this factor may cause - generally this refers to the amount of displacement which the factor could cause and under what, if any, circumstances the rule would be CONCLUSIVELY ABSOLUTE, thereby forcing the APM to divert to an early conclusion and abridging all other factors
  - special case commentary or text justifying why this custom factor was created and its purpose.

  (3) Classification of factor as one of the following:

  - significant only for this case
  - potentially significant for limited additional cases
  - potentially significant for all cases.

****

CODE: FS/APM/INFOREQ NAME: Remaining Information Requirements

1. Enter any additional information required to run the APM

****

CODE: FS/APM/PRERUN NAME: APM Pre-run

1. Generate a PAR TABLE with all Stock and Custom factors which it is intended to include in the pending run of the APM. This table will indicate for each factor its matrix weighting and its 'at-par' value for a default finding of +1 tier displacement.

****

CODE: FS/APM/MATRIX NAME: Assign Factors to Matrix / Run APM

1. In order to assign a given factor to its respective block in the APM matrix, it must be classified in two ways as set out below:
   (1) According to the APM definitions of ABSOLUTE, RELATIVE and MINIMAL. For all stock factors this classification is already made by the stock (default) matrix.
   (2) As one of TECHNICAL, OPERATIONAL or ECONOMIC.
2. Place the factor in the matrix. See TABLE 5.4.
3. Each application placement factor, as derived from a placement driver, can impact one or more of the following attributes of the application: tier placement, service model and target system within a tier.
4. For each factor, the required information must be obtained and input and it must first be determined if the factor applies (this is dealt with below). Where the factor applies, its constituent rule (see Appendix B for the rules for each Stock APM factor) will be invoked where this is possible and will result in one of the following outcomes:

   - result is unclear or even undefined (in some cases this may invoke another residual or remedial factor)
   - result is a displacement finding
   - result is a CONCLUSIVELY ABSOLUTE finding which terminates the consideration of all other factors and makes a final and irrevocable placement of the application.

For example, Factor TA-1 in Appendix B addresses the issue of workspace sharing. The rule for this factor will produce either an undefined result or an Effect ('E' Value) of between 0 and 3. If a value of 2 is produced in a given case this would be multiplied by the weighting value of the block in which this factor falls (here, the value is 9 rendering a positive displacement vote of +18 for this particular factor). Similarly, each other factor within this block will have its rule-generated E Value multiplied by the block weighting factor of 9. The same procedure is followed for each other block (using the weighting value for each such block) with the result that each *relevant* factor for which its rule was successfully executed produces a score of:

- a negative integer for down-tier votes
- zero for neutral votes
- a positive integer for up-tier votes.

Note that not all factors may be relevant to all cases. If the factor is ABSOLUTE it will either specify or 'de-specify' a given tier. Where a tier is specified, subtraction of the value of TIER(DAL) from this tier will usually render the correct displacement value. For example, if a given factor recommends TIER 4 and TIER(DAL) is 2, then the displacement value (E Value) is +2. This is IMPLIED VECTORING since the E Value must be determined by comparing the selected tier to TIER(DAL). Where a tier is de-specified, no action normally results except in the special case where the de-specified tier is actually TIER(DAL). In such a case, it may be desirable to institute either of the following two sets of rules:

- a reserve displacement value to determine whether such a rejection of the home tier should result in up-tier or down-tier movement – this value would be specific to the factor in question; or
- a universal reserve displacement value which the analyst could set at the outset of a given run, which would over-ride any factor-specific reserve displacement values.

If the factor is RELATIVE or MINIMAL in its displacement strength then DIRECT VECTORING can be employed. The rule will normally not determine a specific choice of tier but will directly determine the tier displacement, often using the function MU>> (or MD>>) which causes placement to move up-tier (or down-tier) until a given logical or other mathematical condition is satisfied. The number of tiers moved until the condition is fulfilled is recorded as the E Value in this case.

Thus, the APM algorithm would be to:

- move to first/next block
    - move to first/next factor
    - fetch required data
    - execute rule and determine E Value
    - multiply E Value by block weight and store result
    - continue to next factor or next block
- produce PAR TABLE for factors whose rule were actually executed successfully and for which weighted E Values have been calculated and stored
- find the sum of the weighted E Values
- compare par score and actual score to determine the recommended placement (movement up-tier or down-tier from DAL or else remain at DAL)

- document results and list factors which were not used plus those which were employed but whose rules either failed to fire or produced an undefined or unusable result.

****

CODE: FS/APM/DISPL NAME: Place Application

1. The result of MATRIX (at the whole-matrix level) will almost always be a non-zero integer of some considerable magnitude. A par or default measure is required to permit interpretation of this number. It must also be recalled, however, that with the potential for entry of custom factors which are unique to a given case no two runs of the APM will necessarily produce exactly the same par result. See Chapter 5 for the APM matrix

****

CODE: FS/APM/VERVAL NAME: Verify and Validate APM Factors

1. Verify operation of each step in the APM (from APMASSUM to DISPL inclusive). Review measures include:

- review of application and user group taxonomy
- verification of APM correct logical functioning
- review of key technical, operational and economic issues
- blockage review (does anything block recommended placement?)

****

CODE: FS/APM/EXTSIT NAME: Assess Placement Finding

1. Review information on the existing situation of the workgroup in light of the APM findings. Recall and review any variable or data set which will assist with this process.

****

CODE: FS/APM/REVIEW NAME: Review Deployment Strategy

1. Review deployment strategy and change it as required by invoking DEPLSTR once again or simply modifying variables and data set fields from this module.

****

CODE: FS/EMIT/COSTR NAME: Set Costing Strategy

1. Final review of costing assumptions and COB records.

****

CODE: FS/EMIT/TWI NAME: Total Work Items

1. Collect information on the total cost to operate the workgroup's organizational unit for one year.
2. Find the value of WI(A)/WI(T) as discussed in previous modules.
3. Determine the amount of current cost which is directly attributable to the information management to be accomplished by the proposed application by multiplying the above two results.

****

CODE: FS/EMIT/WICOST   NAME: Cost Per Work Item

1. Use the results of the TWI module to calculate the cost per WI by finding the Sigma of all classes and all members of such classes within the WI(A) universe.

****

CODE: FS/EMIT/TACCOST   NAME: Tactical Costs

1. Review each qualifying record in the COB dataset and accept, modify or reject it as a valid record for the candidate currently under consideration. Of course, candidates which are not open systems will have generally lower scores under this analysis.

2. Repeat for all candidates being considered.

****

CODE: FS/EMIT/STRCOST   NAME: Strategic Costs

1. See Tactical Costs.

****

CODE: FS/EMIT/TACOPP   NAME: Tactical Opportunities

1. See Tactical Costs.

****

CODE: FS/EMIT/STROPP   NAME: Strategic Opportunities

1. See Tactical Costs.

****

CODE: FS/EMIT/TACBEN   NAME: Tactical Benefits

1. See Tactical Costs.

****

CODE: FS/EMIT/STRBEN   NAME: Strategic Benefits

1. See Tactical Costs.

****

CODE: FS/EMIT/ESTM1   NAME: First Formal Cost Estimate

1. Include all TACOST-approved COB records for the selected candidate application, system or application/system.

2. Select and factor STRCOST-approved COB records using only YR 0 or using YR 0 plus YR 1.

****

CODE: FS/OTHR/BESTAPPR   NAME: Best Approach

1. Final review and potential for modification of: net dollar finding of COB, fulfilment of OBJECT records, PLCMT result, CONFIG, ACQ and VENDOR.

****

CODE: FS/OTHR/CAP2 NAME: Second-generation Capital Request
1. Collect information and produce structured document for formal approval of the application to exit the Feasibility phase. At this stage, the information probably should be organized in terms of: business issues and requirements, operational issues, economics technology issues, findings, recommendations and implementation approach.

****

CODE: BC/SFM/CNFG NAME: Final Configuration
1. Increase precision of application/system configuration data and make any minor final configuration changes as required.
2. Improve the cost estimates as required and permit any resulting modifications of the COB data.
3. Where desired, attach actual product names (as examples) to the various layers within the triple-stack SFM model.

****

CODE: BC/SFM/ACQPROC NAME: Acquisition Procedure
1. Finalize and hone choice of acquisition procedure and establish a pro-forma implementation schedule.
2. Change COB records as required.

****

CODE: BC/SFM/CNDLST NAME: List of Candidate Solutions
1. Produce a list of applications, systems or application/system combinations which were considered as candidates during the analysis with or without cost and OBJECT record fulfilment data.

****

CODE: BC/EMIT/ESTM2 NAME: High Confidence Cost Estimate
1. Review ESTM1 module results and modify as required by the results of the CAP2, BESTAPPR, CNFG and ACQPROC modules.

****

CODE: BC/EMIT/CAP3 NAME: Final Draft Capital Request Document
1. Place all required data in format necessary to obtain capital approval to proceed with the project.

****

CODE: BC/EMIT/BUDGET NAME: Create/change Project Budget
1. Based on the COB profiles, the intended acquisition strategy and the previous ESTM1, ESTM2, CAP1, CAP2 and CAP3 modules, set the project budget.

Index